# STUDIA MATHEMATICA 4
# UNISA 1981

# Myocybernetic control models of skeletal muscle

## Characteristics and applications

HERBERT HATZE
*Professor of Biomechanics*
*University of Vienna, Vienna, Austria*

UNIVERSITY OF SOUTH AFRICA
PRETORIA

© 1981 University of South Africa
All rights reserved

**ISBN 0 86981 216 5**

Typesetting and printing by
  Gutenberg Book Printers
  Pretoria

Published by the
  University of South Africa
  Muckleneuk, Pretoria

*To my wife*
   *ROSWITHA*
*and children*
   *GUNTER, ROLAND, and SANDRA*

# Contents

| | | |
|---|---|---|
| PREFACE | | ix |
| CHAPTER 1: | HISTORICAL BACKGROUND | 1 |
| CHAPTER 2: | STRUCTURAL ELEMENTS OF SKELETAL MUSCLE | 16 |
| CHAPTER 3: | FUNCTIONAL CHARACTERISTICS AND MODELS OF THE MUSCLE ELEMENTS | 22 |
| 3.1 | Functional characteristics and models of the parallel elastic elements | 22 |
| 3.2 | Functional characteristics and models of the series elastic elements | 26 |
| 3.3 | Functional characteristics and model of the contractile element | 28 |
| *3.3.1* | *The excitation dynamics* | 31 |
| *3.3.2* | *The contraction dynamics* | 42 |
| CHAPTER 4: | A GLOBAL MYOCYBERNETIC CONTROL MODEL OF SKELETAL MUSCLE | 51 |
| 4.1 | The model of the distributed system | 51 |
| 4.2 | The model of the lumped system | 54 |
| 4.3 | The myodynamics of fusiform muscles | 56 |
| 4.4 | The myodynamics of penniform muscles | 67 |
| CHAPTER 5: | METHODS FOR MODEL PARAMETER ESTIMATION | 71 |
| 5.1 | Morphometric methods | 74 |
| 5.2 | Myodynamic methods | 79 |

| | | |
|---|---|---|
| 5.2.1 | Observations of maximum steady-state isometric torque outputs at various muscle lengths | 79 |
| 5.2.2 | Observations of dynamic isometric torque outputs at a fixed muscle length | 86 |
| 5.2.2.1 | Observations of quasi-stationary torque outputs at various activation levels | 90 |
| 5.2.2.2 | Observations of linearly increasing torque outputs | 91 |
| 5.3 | Myocybernetic methods | 93 |

| | | |
|---|---|---|
| CHAPTER 6: | **MODEL VALIDATION AND COMPUTER EXPERIMENTS** | 96 |
| 6.1 | Computerized myocybernetic models | 98 |
| 6.2 | Computer experiments on model muscle fibres | 100 |
| 6.2.1 | Simulation of twitch responses | 100 |
| 6.2.2 | Simulation of multi-stimulative responses | 107 |
| 6.3 | Computer experiments on model muscles | 119 |

| | | |
|---|---|---|
| CHAPTER 7: | **APPLICATIONS** | 122 |
| 7.1 | Hominoid dynamics | 122 |
| 7.1.1 | Definitions | 123 |
| 7.1.2 | The model of the executor subsystem | 125 |
| 7.1.3 | The model of the myoactuator subsystem | 128 |
| 7.1.4 | The model of the complete neuro-musculoskeletal system | 130 |
| 7.1.5 | Simulation of the long-jump take-off phase | 130 |
| 7.2 | Foundations of myocybernetics | 136 |
| 7.2.1 | Myoenergetics | 136 |
| 7.2.2 | Formulation of the general myocybernetic performance criterion and the myocybernetic optimal control problem | 141 |
| 7.2.3 | Examples of optimal myocybernetic control modes | 143 |
| 7.2.4 | A principle of optimal grading sensitivity | 146 |

**APPENDICES**

| | | |
|---|---|---|
| A1. | Derivation of the state equations of the global myocybernetic control model | 150 |
| A2. | Derivation of the model equations for penniform muscles | 165 |
| A3. | Computer program ELPEST for estimating myodynamic parameter values | 169 |
| A4. | Computer program MYOSIM for simulating myocybernetic models | 179 |
| BIBLIOGRAPHY | | 201 |
| INDEX | | 213 |

# Preface

The impetus towards the writing of this monograph had two sources. In 1975, Prof. H.S.P. Grässer of the Department of Applied Mathematics of the University of South Africa invited me, on behalf of the University, to write a comprehensive exposition of my work on myocybernetic control models of skeletal muscle, investigations which had been described in several publications and in my Ph.D. thesis. At that stage, however, I felt that the subject of myocybernetics had not matured sufficiently to warrant a comprehensive treatise.

Further work on the subject progressed reasonably well and certain basic concepts began to emerge. This new development aroused interest among muscle researchers and bioengineers, resulting in an invitation to present an overview of the field in the form of a keynote address delivered to the participants of the Engineering Foundation Conference on 'Biomechanics of Movement' which was held in Henniker, N.H., in June 1979.

At this conference many participants suggested that, in their view, the urgent need had arisen for presenting the volume of material on the myocybernetics of skeletal muscle in the form of a comprehensive treatise. These suggestions provided the second incentive to write this monograph.

Recent extensions and generalizations of previously published material which have not appeared elsewhere are presented here for the first time. In this sense, the present monograph represents the most up-to-date account of the subject area.

The book has been written with the needs of muscle biologists, bioengineers, and sport scientists in mind. All of these researchers require a profound understanding, not only of the microbiological structure and function of skeletal muscle, but also of its behaviour under various neural control modes. Perhaps the most appropriate way of describing this control behaviour is to create a myocybernetic control model of skeletal muscle, and then simulate on the computer its responses to various neural control inputs. If the model is adequate, its responses will closely mimic those of the living muscle, for a given set of muscle-specific parameters and neural control inputs. The model can then also be used to predict muscular responses to control modes which are difficult or impossible to generate experimentally.

A thorough literature search undertaken in the late 1960s convinced the author that none of the models proposed up to that time could justifiably be called a myocybernetic control model of skeletal muscle. Most modelling attempts had concentrated on the functional behaviour of the myo-structures with special reference to the sliding filament theory, but failed to incorporate the intricacies of the excitation and recruitment dynamics. However, a vast volume of experimental data was available which enabled the author over the past ten years successively to construct, refine and test a model which could be regarded a true control model of skeletal muscle.

The present monograph is designed to give a step-by-step exposition of this development. Detailed mathematical derivations have been transferred to the various appendices in order to make the mathematical treatment presented in the main text accessible to the average biologist with a moderate background in mathematics.

The book is organized into seven chapters and a number of appendices. Chapter one provides a historical background of theories and models of muscular contraction. In Chapter two we identify the structural elements of skeletal muscle. Chapter three is devoted to the description and modelling of the functional characteristics of the muscle elements, while in Chapter four these submodels are combined in the global myocybernetic control model. In Chapter five we describe methods for determining model parameters from measurements on living muscle, and Chapter six contains a comparison between simulation and corresponding experimental results. Finally, Chapter seven gives an exposition of various applications of the model.

Since simulation constitutes one of the main features of the present approach, computer programs are provided which enable the reader to perform his own model simulations and parameter determinations. The computer programs are written in ANSI FORTRAN IV and are listed in the appendices.

In concluding this preface, I wish to express my gratitude towards Prof. H.S.P. Grässer and the University of South Africa for inviting me to write this monograph. For valuable discussions, thanks are also due to Prof. E. Henneman of Harvard University, Prof. D.H. Jacobson and Dr D.H. Martin of the National Research Institute for Mathematical

Sciences of the CSIR, and Prof. R.B. Stein of the University of Alberta, while I am grateful to Mr F.R. Baudert for carefully editing the text.

*Pretoria, November 1979*

# CHAPTER 1

# Historical background

The understanding by the reader of the development of the myocybernetic control model of mammalian skeletal muscle to be presented in this monograph will be greatly enhanced by a knowledge of the history of muscle models and contraction theories. For this reason, the present chapter will be devoted to a brief description of the historical development which has led to the muscle models in use today.

Throughout this monograph the term 'model' should be understood to mean the abstract (verbal, graphical, or mathematical) description of selected attributes of a real entity (the skeletal muscle).

The phenomenon of muscular contraction as the propellant of animal motion has fascinated man ever since he began to view nature analytically. The earliest written accounts on the subject date back to the fifth century B.C. The Hippocratic collection of writings on medicine and its philosophy, which includes ideas of even earlier times, consists of various treatises of the corpus (some sixty in number), representing sometimes even conflicting views (Needham, 1971).

It is interesting to note that in these early Greek works the tendons were confused with nerves. Littré (1839, p.183) describes this view: 'The bones give a body support, straightness and form; the nerves give the power of bending, contraction and extension; the flesh and the skin bind the whole together and confer arrangement on it; the blood-vessels spread throughout the body, supply breath and flux and initiate movement'.

Aristotle (384 to 322 B.C.), influenced by these ideas, also considered animal motion to be caused directly by the power of the nerves and the spirit. He says: 'The movements of animals may be compared with those of automatic puppets which are set going on the occasion of a tiny movement; the levers are released, and strike the twisted strings against one another ... Animals have parts of a similar kind, their organs, the sinewy tendons to wit and the bones; the bones are like the wooden levers in the automaton, and the iron; the tendons are like the strings, for when these are tightened or released movement begins ... Now experience shows us that animals do both possess connatural spirit and derive power from it ... And this spirit appears to stand to the soul-centre or original in a relation analogous to that between the point in a joint which moves, and that which is unmoved. Now since this centre is for some animals in the heart, in the rest in a part analogous with the heart, we further see the reason for the connatural spirit being situated where it actually is found ... We see that it is well disposed to excite movement and to exert power; and the functions of movement are thrusting and pulling. Accordingly the organ of movement must be capable of expanding and contracting; and this is precisely the characteristic of spirit. It contracts and expands naturally and so is able to pull and to thrust from one and the same cause, exhibiting gravity compared with the fiery element, and levity compared with the opposites of fire' - see Farquharson (1912, p. 703a).

It was not until the early third century B.C. that the muscles were mentioned explicitly as being the direct cause of animal motion. This hint is contained in the works of Herophilus of the Alexandrian school, which are known to us mainly through the writings of Galen (Kühn, 1821).

At the same time, a younger contemporary of Herophilus, Erasistratus, recognised that muscles are contracting structures. He postulated the first theory of muscular contraction (at least the first we know of). 'Erasistratus says that the muscles, if they are filled with pneuma (breath), increase in breadth but diminish in length, and for this reason are contracted' (see Kühn, 1821, p. 429).

Rufus of Ephesus (early second century B.C.) advanced the knowledge on the subject further when he studied dissected muscles and recognised them as tissues built on a particular pattern.

Possibly the greatest contribution ever made to muscle physiology is that by Galen (129 to 201 A.D.). He realized that muscle can only contract or relax. He also made it clear that the rich variety of possible limb movements is the result of a complicated system of differently positioned muscles. A further important contribution of Galen was his successful demonstration of the communication of the muscles with the spinal cord (see Daremberg, 1854), which laid the foundations for an understanding of the control of muscle by the central nervous system. Galen's conceptions were to dominate the field for the next 1300 years.

In the sixteenth century, the first attempts were made to explain the mechanism of muscular contraction. Although the works of Vesalius, Fallopius and his pupil Fabricius (later the teacher of William Harvey) were not directly concerned with the explanation of the mechanism of muscular shortening, they nevertheless refined the concepts of the morphology of muscle which were so vital for the subsequent contraction theories.

Surprisingly, the next step towards a deeper understanding of the phenomenon of muscular contraction was made by Descartes, better known as a mathematician and philosopher than as a physiologist. It may be appropriate to devote a little space to Descartes's ideas (1646) on the subject: 'Now according as these spirits enter thus into the concavities of the brain, they pass thence into the pores of the substance, and from these pores into the nerves; where according as they enter or even only as they tend to enter more or less into the one rather than into the others, they have the power to change the shape of the muscles in which these nerves are inserted, and by this means to make all the limbs move. Just as you can have seen in the grottoes and foundations in the gardens of our Kings, that the same single force by which the water moves, coming out from its source, is enough to move diverse machines and even to make them play instruments or to pronounce words according to the different disposition of the tubes which conduct it' (see Needham, 1971, p.14). This is possibly the first attempt to build a conceptual model of muscle. Descartes refers here to the 'hydraulic gardens' of the seventeenth century and their automatic puppet-plays. He compares the nerves with the tubes of the fountain machines, the muscles with the 'diverse engines and springs operating in these machines' and the 'spirit' with the water running through the fountains. This mechanistic view of muscular contraction is not surprising: Descartes was a product of his time, a time during which the foundations of the science of mechanics were laid by Galileo and Newton.

Interest in research on muscle was greatly on the increase at this time. William Croone published his first work on muscular movement, 'De ratione motus musculorum', in 1664. In the same year appeared the treatise 'De musculis et glandulis observationum specimen' by the Danish anatomist Nicholas Stensen. So striking are his descriptions of the motor fibres and fibrils of the muscle that it appears likely that he had made microscopic observations. He also made it clear that not the tendons but only the fibres are responsible for the contraction.

At this time, experiments carried out by Jan Swammerdam and Jonathan Goddard proved beyond doubt that the contracting muscle does not increase its volume as had been believed earlier. This finding had a strong bearing on Croone's theory of muscle contraction which was published in 1675. In his theory, he proposes that the muscle fibre consists of many tiny globules, and that contraction is brought about by 'muscle juice' filling these globules. This was also the physiologist Alfonso Borelli's picture of the muscle structure (see Wilson, 1961). Another concept to explain the origin of muscle contraction was proposed by John Mayow in 1674. Briefly, his theory was of a biochemical nature and assumed that the contraction of the muscle was due to contortion of the fibrils 'caused by nitro-aerial particles set in motion and even pretty intensely warmed in the motor parts'. This concept closely resembles today's accepted theory of the aerobic processes taking place in muscle. Mayow's contribution was, however, ignored by his contemporaries and only revived about a hundred years later.

This marks the end of the earlier history of the subject. A rather inactive period followed, and the next significant advances were not made until 1870. In reviewing developments made since then, only the more important theories of contraction and their associated models of muscle will be considered. In discussing these theories and models, it will be necessary to use certain subject-specific terms, with which the reader will be assumed to be familiar. The terms used in the structural description of skeletal muscle (Chapter 2) will, however, be defined when they first appear.

In 1873, Engelmann (1873) advanced the imbibition theory, which postulates a passage of water from the isotropic to the anisotropic bands, followed by imbibition of this water by a molecular complex which is situated along the axis of the fibre. These elements then change into a spherical form, thus causing a decrease in their length. In addition, Engelmann conceived muscle as a heat engine. In order to justify this assumption he had to assume that some infinitesimal parts reach a temperature of $140°$ C on contraction. This high temperature, he thought, is the result of some unknown chemical reactions which provide the energy for the contraction process. Relaxation, and thus lengthening of the shortened fibres, would be accomplished by elastic forces of parts passively moved during shortening, once the heat-yielding reactions had ceased.

The imbibition theory, stripped of its thermodynamic aspects, appears to have been very attractive to subsequent investigators. As the reason for the imbibition, Pauli (1912) substituted increasing electrostatic forces for the rise of temperature. These forces, he hypothesised, were due to a change of the contractile proteins from an iso-electric state into a cationic state, as a result of the increase in the hydrogen ion concentration on the formation of lactic acid. As regards the source of the imbibition fluid, McDougall (1897) and Meigs (1908) expressed the opinion that fluid passed from the sarcoplasm into the sarcomere, the cause of this flow being the breaking up of the inogen molecule into a number of smaller parts.

The imbibition theory was rendered unlikely when Weber (1927) emphasised that the enzyme system is inhibited at a pH value of about 6.0 and thus lactic acid formation is self-hindered. He also showed that isoelectric protein does not exist in the form of uncharged molecules but rather as zwitter ions.

Another interesting development was the surface tension theory. Bernstein (1915) ascribed the tension produced in the fibre to the increase of surface tension of tiny ellipsoids, connected by elastic structures, and making up the fibrils. Lactic acid formation was thought to be the trigger for the change of these ellipsoids into spheres (this change of shape decreasing their surface area). Initially, there seemed to be some experimental support for this theory, since the maximal muscle twitch tension increases with decreasing temperature, and the temperature coefficient of the surface tension changes is negative. Hill (1923) was at first in favour of this theory, but soon rejected it when he had made new calculations and taken into account fresh experimental evidence.

In 1924 a theory was proposed by Meyerhof (1924) which regards the dehydration of protein as the source of the contraction force. This theory was criticised one year later by Wöhlisch (1925). Wöhlisch also rejected the imbibition theory on the grounds that both theories were based on assumptions (regarding the supposed effects of lactic acid) which were incompatible with protein properties known at that time. He advanced his own theory which assumes the release of a special substance on stimulation. This substance would cause a decrease of the colloidosmotic pressure in so-called ultrafibrils (which Wöhlisch pictured as substructures of the fibrils with elastic and semi-permeable walls), which under the influence of their stretched elastic membranes would start to contract. On removal of the substance the muscle would return to its original length.

This theory does not seem to have produced much impact, and was soon followed by another involving alterations in the electrical charges of the protein chains. The latter theory, which is due to K.H. Meyer (1929), postulates that there exists an alternating arrangement of acid and basic groups in fibrous protein chains. Meyer regarded these chains as made up of zwitter ions and having a length of about 400 Å and a width of about 4 Å. De-ionisation of the molecular groups in the chains would lead to folding, while ionisation would reverse the situation. He calculated an average utilisation of only 5% of the total muscle protein and a maximum shortening to 85% of the rest length of the muscle (intact muscle shortens maximally to 60%). The latter point was heavily critisised by Weber (1930) on the grounds that in the non-stimulated muscle about two-thirds of the positively-charged groups remain ionised so that the chains could not be fully stretched by repulsion between negatively-charged groups. Weber only admitted a possible shortening of 5 – 30%.

It becomes apparent from this brief discussion of the contraction theories of the period from 1870 to about 1939 that the investigators were concentrating more and more on the fine substructure of the muscle material. Advances in biochemistry prompted theories of contraction which reflected these new views. This is even more true for the period that follows and which is marked by the discovery of the ATPase activity of myosin by Engelhardt & Lyubimova (1939). This discovery and subsequent developments, as well as other aspects of the contraction theories discussed earlier, are brilliantly described by D.M. Needham (1971) and make interesting reading.

The question may be asked why so much space is being devoted to the description of theories of muscle contraction. In fact, contraction theories generally lead to the construction of models of muscle (even if these are not called by that name), which may then be used to test the validity of the underlying theories. This is true to an even greater extent for the more advanced and complex theories which will now be discussed.

Since the contractile proteins have a highly specific configuration and myosin contains a high proportion of polar groups, it was very unlikely in the opinion of W.T. Astbury (1950) that the chains would change into a random configuration (accompanied by an entropy change). It was thus more likely that the postulated folding of the protein chains in contraction of the muscle would be a result of internal energy changes and not changes in entropy. This hypothesis received support from experiments by H.J. Woods (1946) on the thermodynamic behaviour of oriented myosin strips, the experiments showing that the contribution due to changes in entropy was indeed negligible.

It has already been mentioned that the discovery, by Engelhardt and his collaborators, of the ATPase activity of actomyosin gave a new impetus to the research on muscle contraction. Engelhardt himself (Engelhardt and Lyubimova, 1942) used his findings to propose a possible mechanism of contraction. He imagined the following successive events to happen: ATP, which is bound in resting muscle to some protein complex, combines with myosin on stimulation; this initiates some unknown processes inside the molecular myosin structure, and contraction ensues; during this process, ATP was dephosphorylated (thus yielding the energy for the contraction) and ADP, which has a lower affinity for myosin than ATP, dissociates from the myosin, which returns to its original configuration.

Sandow (1946) took a somewhat different view of the issue. He suggested that the effects observed experimentally by Engelhardt *et al.*

were really an expression of latency relaxation. He was of the opinion that it was the energy release itself which followed the ATP breakdown that caused the activation of myosin for contraction. He termed this scheme contraction-energisation.

A different interpretation of experimental results available at that time was given by Varga (1946). In his theory of contraction he postulated that there were only two possible states of the actomyosin particle: a completely contracted and a completely relaxed state. A change between the states would constitute a reversible chemical reaction, the equilibrium point of which depended on the temperature, and absorption of heat would occur in the contraction cycle.

A modification of Meyer's theory (mentioned earlier) was suggested by Riseman & Kirkwood (1948). They proposed a mechanism in which the energy released by the breakdown of ATP would be stored in the myosin chain in the form of electrostatic potential energy. This would lead to an extension of the myosin chain, in the resting state, due to repulsion of neighbouring negative charges. Contraction would result when the negative charges were removed as a consequence of the dephosphorylation of the myosin molecules.

A strong argument against this theory by Pryor (1950) was that the acting electrostatic forces should cause extensions or contractions in all directions and not merely in the direction of the fibre axis.

A group of theories, published in the 1950s (Weber & Weber, 1950, Pryor, 1950, to mention but a few), was based on the assumption that the contractile energy came from an increase in entropy by chemical decomposition. Pryor's theory was refuted when A.V. Hill (1953) showed that active muscle did not behave in the way this theory proposed.

Weber & Weber (1951) reconsidered the question of the energy provision when experimental evidence made the entropy theories unlikely. Their new conception was that the energy derived from the hydrolysis of ATP was directly responsible for the contraction. The temperature dependence of the contraction force was thought to be a result of the temperature dependence of the speed of the chemical reaction.

A theory involving actin was proposed by Mommaerts (1951). He pictured resting muscle as containing myosin and G-actin in combination. On stimulation, actin would react with a stoichiometric amount of ATP and dephosphorylate this during polymerisation. The result would be F-actomyosin, which would be the contracting agent. Rephosphorylation of ADP by a phosphate donor would occur during relaxation. Mommaerts himself admitted later that the stoichiometric relations underlying his theory represented only a limiting case.

Possibly the greatest break-through in the field of muscle research since Galen was made by H.E. Huxley (1953) when he presented his sliding-filament theory. This theory is generally accepted today as the basis of the actual shortening mechanism. As experimental evidence for the correctness of the new theory mounted, the theories based on the longitudinal folding of the contractile proteins were quickly discarded. D. Needham (1971) has described the history of the discovery of the sliding mechanism in a very entertaining way. She relates in detail how X-ray studies slowly convinced Huxley that the molecular structures were in fact sliding past each other during the contractile process, and not folding, as had been believed previously.

It is, unfortunately, not possible in this context to go into the details of the sliding mechanism. Thus, familiarity of the reader with this concept will be assumed when discussing contraction theories based on filament sliding. Such familiarity will, however, not be presupposed in the following chapters, when the mathematical model of skeletal muscle is treated.

After this stage of research had been reached, theories centred around the question of *how* the sliding was produced and where the energy was derived from. Again, only the more significant theories will be mentioned.

An interesting study, published by A.F. Huxley (1957), dealt with the connecting mechanism between the myosin and the actin filaments. It was actually more of a hypothesis than a theory since the existence of links between the two types of filaments had not been established at that time. Huxley hypothesised that side-pieces, bearing active sites which were oscillating in the direction of the filament axis in a Brownian motion, were situated along the myosin molecule. Each active site could react with a suitable site on the actin molecule if the sites approached each other closely enough. Huxley pictured the active myosin site as oscillating between two springs (on a molecular scale, of course). If the stretched spring made contact with an actin site, it would pull the actin alongside the myosin, thus producing the shortening. ATP was assigned the role of breaking the actomyosin link and, simultaneously, restoring the energy in the spring-like contraction mechanism. The rate constants, for the making and breaking respectively of the actomyosin connection, were postulated as depending on the distance of the actin site from the equilibrium position of the oscillating myosin site. The mathematical model associated with this hypothesis produced reasonable results when specific values were assigned to the rate contants.

Huxley's model, although exceedingly successful at the time of its creation, has lost its attractiveness in the light of more recent experimental evidence. Firstly, the results of Bárány (1967) concerning the ATPase activity of myosin clearly indicate that the rate limiting factor is the ATP splitting. This contradicts Huxley's assumption that one of his rate constants (the constant f) is responsible for the limitation of the reaction rate. Secondly, recent work of Lowey et al. (1969) on the substructure of the myosin molecule makes Huxley's concept of the form of the actomyosin connection extremely unlikely. Lowey *et al.* have clearly demonstrated that the side-pieces emerging from the myosin backbone consist of two (more or less) identical rods of light-meromyosin (LMM) which are connected to two pieces of heavy-meromyosin subfragment 2 (HMM-S2). These, in turn, each bear one globular head (HMM-S1). With this type of structural configuration a contractile mechanism in the form proposed by Huxley would not be possible.

In the same year in which A.F. Huxley published his conception of the contractile machinery (i.e. in 1957), H.E. Huxley (1957) provided electronmicroscopic proof for the existence of structures linking the myosin and actin molecules. These structures were termed cross-bridges and Huxley suggested in his paper (influenced by works of A.G. Szent-Györgyi) that these bridges consisted of the heavy meromyosin parts of the total myosin molecule. This would mean that the location of the ATPase activity and the site of the contact with actin had been found. This was a tremendous step forward, and all subsequent theories had to take this finding into account.

A theory that had a considerable impact was that of R.E. Davies (1963). Davies, like others, concentrated mainly on the events happening in the cross-bridges. Basically, the cross-bridges are imagined to exist in two different states. In the inactive state the bridges consist of extended polypeptide chains with fixed negative charges at their bases. One ATP molecule is bound to the top of each bridge in a way which results in a negative unit net charge. The repulsion between the two negative charges (base and top) keeps the bridge extended. If the muscle is stimulated, calcium moves in from the sarcoplasmic reticulum, and one calcium ion (2 positive unit charges) provides a link between the $ATP^-$ on top of the cross-bridge and the $ADP^-$ situated on the actin filament. Since now the negative charge on top of the bridge is neutralised, repulsion is no longer present and the extended polypeptide chain transforms into a helix-coil. This process is supposed to produce the shortening. A few more steps (involving the dephosphorylation of the ATP and the abolition of the calcium bond) follow and then the cycle starts again.

When more became known about the micro-structure of the molecular filaments, the pillars on which Davies's theory was built began to crumble. What perhaps above all rendered the theory highly unlikely, was the fact that the role which calcium plays in the contractile process was clarified. Ebashi and Endo (1968) could show that calcium acts indirectly by combining with the troponin-tropomyosin complex, thereby removing the inhibition of the actin-myosin interaction. This was not the only weak point in the theory. Mommaerts, in his 1969 review on muscular contraction, has pinpointed a series of inconsistencies in Davies's theory.

More recently developed theories are those of Loewy (1968), Ebashi, Kodama & Ebashi (1968), Perry & Cotterill (1965) and Bendall (1969). All these papers, like the work of Davies, are concerned with the succession of chemical events taking place between the cross-bridges and the actin molecule with its associated troponin-tropomyosin complex, and will not be discussed in detail.

A contraction hypothesis worth mentioning is that developed by Elliot, Rome & Spencer (1970). These authors base their views on the fact that the volume of the muscle fibre does not change when it contracts. This implies that the axial distance between the filaments in the lattice must increase on shortening. In the above hypothesis, however, the arguments are reversed: it is envisaged that the interaction of myosin with actin increases the negative charge on the actin molecule, thus increasing the repulsive coulombic forces between the filament groups. This would then necessarily lead to shortening (by virtue of the constant volume relation). The creators of this hypothesis have suggested several methods for investigating the postulated phenomena.

H.E. Huxley (1969) proposed a possible mode of operation of the globular heads of the cross-bridges, which had a marked influence on the writer's own theory of muscular contraction. Huxley conceives minute sliding movements of the globular heads relative to one another to be the cause of the contractile motion. This view is similar to that of the writer himself regarding the ultimate contractile mechanism.

The writer's theory (Hatze, 1973) is based on the assumption that the chemical energy available for the contraction process in muscle is first converted into electrical energy and then from electrical energy into mechanical work and heat. The sliding filament mechanism is incorporated into the theory as far as the relative movement of the actin and myosin filaments is concerned. The anatomical model adopted is basically that described by H.E. Huxley (1969) and by Lowey *et al.* (1969) as regards the globular heads of the heavy meromyosinmolecule. Briefly, the hypothetical contraction process is conceived to take place as follows: on the arrival of a nerve signal at the motor end-plate an action potential is elicited which propagates along the fibre and down the T-tubular system. This action potential then causes the release of $Ca^{2+}$ from the terminal cisternae in the sarcoplasmic reticulum near the Z-line. The calcium ions are now assigned a twofold function: they initiate the hydrolysis of ATP and, simultaneously, are thought to decrease the electric resistance of the troponin-tropomyosin complex where they bind, thus creating 'suitable' actin-binding sites. The energy freed by the hydrolysis of ATP is assumed to be available in the form of a negative charge on the upper part of the mobile HMM-S1 subunit. After the cross-bridge has been lifted from the myosin backbone by electrostatic attraction (due to the negative charge on its top), it attaches to a suitable binding site on the actin molecule. The prevailing potential difference is then thought to produce a flow of charges which, in turn, sets up an oscillating magnetic field. The resulting mechanical force tends to separate the two globular subunits and since the mobile unit is tightly linked to actin, a relative movement between the filaments ensues. As the magnetic field (and hence also the mechanical force) increases, the electric field declines. Since the bond strength is, by hypothesis, a function of the field strength, an instant occurs when the bond breaks and the mobile subunit returns to its equilibrium position (under the influence of intermolecular forces) only to recommence a new cycle. This cyclic activity: attachment, contraction, detachement, return to the equilibrium position, reattachment - continues until the stored electrical energy is almost used up. The cross-bridge is then ready for the reaction with the next ATP molecule, and a new cycle commences.

This contraction hypothesis is certainly highly speculative (although conceivable). However, the merit of the theory should be judged by the unusual capability of the associated mathematical model to predict known muscular phenomena. The agreement between experimental results and the predictions of the model is striking indeed: the force-velocity relation, the tension-length curve, the heat and work rates, the metabolic rates and all known features of the stretched, stimulated muscle (no ATP-splitting, stretching tension higher than isometric tension, etc.) can be deduced from the solutions of the describing set of differential equations.

Up to this point, mainly contraction theories and molecular models of muscle have been considered. There are, however numerous modelling attempts that are concerned only with the functional properties of muscle. Ideally, molecular models should be capable of describing the behaviour of the total muscle if considered in its function as the motor of limb movements. The situation is, however, similar to that in thermodynamics: when the behaviour of the total gas volume is to be described, the individual particle must be assigned certain 'average' properties. This approach is typical for all distributed systems comprising large numbers of individual particles. Thus, when modelling the muscle as the propellant of animal motion, integrated into the link system of the animal, a compromise has to be reached between unmanageable mathematical complexity and oversimplification.

Returning now to the biomechanical models of muscle, only the more significant contributions to this field will be treated. A much preferred concept (especially among engineers) is a three-component model of muscle in which the components are linear. This model exists in two versions: the Voigt model and the Maxwell model (see Fig. 1.1).

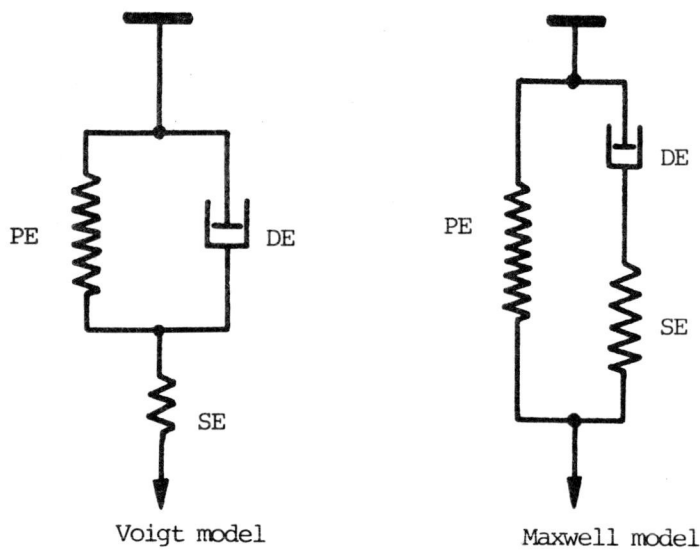

**FIG.1.1:** Two representations of the three-component muscle model. PE, SE and DE denote parallel elastic element, series elastic element and damping element, respectively

Such a model was employed by Williams & Edwin (1970) in the electronic simulation of frog muscle responses to pulse trains at various selected frequencies of stimulation. Although reasonably adequate for this special purpose, the simulator cannot be used to predict any other functional property of muscle such as the force-velocity relation or the length-tension curve.

Another attempt to use the linear three-component model in conjunction with investigations on intrafusal fibres was made by Crowe (1970). He employed the Voigt version of the model to study the responses, to mechanical stretch and fusimotor stimulation, of intrafusal muscle fibres of the mammalian muscle spindle. Crowe postulates that stimulation produces, abruptly, an increase in the moduli of elasticity and a reduction of the unstretched lengths of the elastic elements, as well as an increase in the viscosity of the damping element. On cessation of stimulation, the elements are supposed to revert immediately to their original states.

This model appears to be too unrealistic since the parallel and series elastic elements can be identified with real muscle structures, the properties of which are known to be independent of the degree of stimulation. Moreover, it is well known that the active state of the muscle does not reach its full intensity immediately upon stimulation (Bahler, Fales & Zierler, 1967), nor does it revert to zero abruptly after stimulation has ceased. Other defects of the model have been pointed out by Crowe himself. The model does, however, have its merits as an analogue for some highly specialised contractive situations.

A model of muscle which may be regarded as a control model, is due to Green (1969). Green uses the Voigt version of the three-component representation and makes two components, the parallel spring and the damper, neurally controllable. He also recognises the nonlinearity of all the elements involved, and models the active state as an exponentially declining function of time when treating the isometric twitch. However, it is not clear how the control function $n(t)$, defined by Green as the 'effective number of active axons in the motor nerve', is used to obtain the relationship between peak isometric tension and the frequency of stimulation. A similar argument applies to his second control function m, which is defined as the number of parallel fibres contracting simultaneously. Both functions disappear after they have been defined and never reappear in any of the model equations. Also, Green's assumption that the onset of the active state is immediate can lead to serious errors if situations are considered where the muscle as part of the skeletal system controls rapid motions. In addition, after having defined his nonlinear model, Green converts it back to a linear one (by linearising all components at arbitrarily selected points of their characteristic curves) which restricts the range of applicability of the model to a relatively small portion of the physiological domain. Finally, one would have to criticise the idea that the parallel elastic element is neurally controllable, since it represents (as pointed out above) a real passive structure of the muscle.

A more appropriate mathematical model of mammalian skeletal muscle appears to be one created by Bahler (1968). This model departs from the dashpot idea (the element DE in Fig. 1.1) and replaces it by a so-called contractile element. Since Bahler's model is only valid for muscle lengths of less than 120% of the resting length, the parallel elastic element may be neglected and the model only consists of a contractile element in series with a nonlinear series elastic element. The contractile element is divided into a force generator (the characteristics of which are equal to the length-tension relation), in parallel with an internal load. Bahler realised that it is necessary to separate the passive elements of the muscle (parallel and series elastic element) from the actual contractile element. This conception is much more realistic than the previously discussed models as regards the actual anatomical arrangement. The actual equation used to describe the functional behaviour of the series elastic element is a third-order polynomial. This is in disagreement with the finding of many investigators (e.g. Jewell & Wilkie, 1958) that the load-extension curve of frog and mammalian muscle is approximately exponential (i.e. Bahler's extension-load characteristic should be a logarithmic function). For simulation purposes, however, and for a restricted range, Bahler's approximation may be sufficiently accurate. Another disturbing fact is that no active state function appears in Bahler's model. Thus the model cannot be used for simulations which include transition phases from rest to activity or from activity to rest.

An interesting fact about all the models discussed is that apparently none of them accounts for the peculiar behaviour of the force-velocity relation at positive (i.e. stretching) velocities. Most models adopt Hill's force-velocity equation, which is adequate for negative (i.e. shortening) velocities, but is known *not* to fit the experimental curves for positive velocities. Now, the behaviour of the stimulated muscle and its force production at positive velocities is of cardinal importance with respect to its function in the intact animal body (stretch reflex). Thus all models using Hill's equation for this mode of operation would tend to grossly underestimate the stretching velocity for a given stretching force.

One of the first investigators to include a simple activation control parameter $a$ in a muscle model was Stark (1968). This parameter represents the fraction of stimulated cross-sectional area of the muscle, but does not account for the degree of activation of the stimulated part, nor are the transient features of the excitation process considered. In addition, Stark's control parameter $a$ appears linearly in the model's force output, which necessitates bang-bang (discontinuous) switchings of the muscle force in maximum-effort contractions. However, such a behaviour has never been observed experimentally. Indeed, it is well established that if a resting muscle is stimulated abruptly and maximally, its force output does not immediately attain a maximum (as it would in a bang-bang switching) but rises comparatively slowly in the form of a sigmoid curve. Hence any control model of muscle which predicts a maximal jump in the muscle's force output upon maximal stimulation must be regarded as highly unrealistic.

More recently, FitzHugh (1977) presented a control model of muscle which, as far as the inclusion of the control parameter U is concerned, is an exact copy of Stark's model, with U replacing Stark's control $a$. Hence all comments made above apply equally well to this model. In addition, there are many more inconsistencies and contradictions in FitzHugh's model which are discussed elsewhere (Hatze, 1978a).

Similar remarks apply to the control model of Morecki *et al.* (1975). In this model, the muscle's force output is also multiplied by an 'excitation' variable ū which leads to the same problems as those discussed above.

Bawa *et al.* (1976) recently presented a model of muscle consisting of linear springs and a dashpot in parallel with a force generator. This force generator produces a contractile force of constant magnitude each time a stimulus is received. Immediately afterwards, the force is assumed to decay exponentially. According to the authors, the predictions of this model are in good agreement with experimental observations. This is somewhat surprising since it is well known that all the components used in the model of Bawa *et al.* are, in actual fact, not linear but highly nonlinear, and that the characteristics of the force generator (length-tension relation, excitation and contraction dynamics, etc.) are much more involved than has been accounted for in that model.

With this note we conclude our discussion of the historical development of contraction theories and muscle models, and the associated problems. It has become apparent that a great deal is known today about the molecular mechanism and the phenomenological behaviour of the myostructures. However, it is equally apparent that none of the muscle models discussed, adequately describes the dynamics of the muscular force output in response to the neural control mechanisms motor unit recruitment and stimulation rate. The present monograph is designed to fill this gap. The next chapter will be devoted to a description of the structural elements of skeletal muscle.

## CHAPTER 2

# Structural elements of skeletal muscle

This chapter is intended to provide the transition from the actual anatomical structure of skeletal muscle to the abstract modelling structure. A model is generally considered adequate if it approximates the real situation 'sufficiently closely' for a given purpose. In this investigation it is intended to simulate the functional control behaviour of the muscle as a driving-structure in the link system of the animal, an aim that precludes modelling at the molecular level as irrelevant to the problem at hand and, at the same time, defines the model as one that is mechanical in nature. This, in turn, means that the structural elements of the muscle will have to correspond to mechanical structure elements in the model.

Skeletal muscles usually originate on the skeleton, span one or more joints and insert again into a part of the bone structure. Each muscle (mammalian skeletal muscle is implied) is enclosed in a connective tissue sheath (the epimysium), and, in addition, is held in its correct position in the body by layers of fascia. The muscle is attached to the skeleton by tendons which are bundles of white connective tissue. The interior of the muscle is compartmentalized into longitudinal sections (the fasciculi) which contain up to 150 individual muscle fibres each (Kelly, 1971). These fibre bundles are wrapped within sheaths of connective material, called the perimysium. Each of the fibres contained within the fasciculi is surrounded similarly by the endomysium, the function of which is to support the nerve fibres, capillaries and veins connecting the muscle fibre to the central supply system. Each muscle fibre has its own cell membrane, the sarcolemma. The chemical function of the sarcolemma is very complex, but is of no relevance to the discussion of its mechanical properties.

Muscle fibres do not always run parallel to the force-transmitting tendons, as they do in fusiform muscles. The fibres may be arranged in unipennate, bipennate or multipennate form (see Fig. 2.1).

**FIG. 2.1:** Schematic representation of the unipennate, bipennate and multipennate fibre arrangement in mammalian skeletal muscle

These special arrangements have important mechanical implications, as will be shown in Chapter 3. Although the muscle fibre constitutes a very important unit in the contractile machinery it is not the ultimate structural element in the muscle. Subdividing further, it is found that the fibre contains bundles of fibrils. Each fibril has diameter of about 0.5 microns and comprises the hexagonal array of the protein filaments that are directly reponsible for the contractile process. When the fibrils are investigated under the electron microscope a peculiar structure becomes visible. Certain bands (I, A and H-bands) appear in the picture marking the boundaries of the fibril substructure. The I-bands are centred on the so-called Z-lines from where the actin filaments emanate. To the left and right of an I-band appear the A-bands, which are due to the myosin filaments. In the middle of each A-band lies a lighter region, the so-called H-band. Overlap of the actin and myosin filaments occurs only in the region A minus H. As one moves along the fibril there is a cyclic repetition of the I-band/H-band pattern and the distance between two successive Z-lines defines the length of one sarcomere. The corresponding molecular substructure, including approximate dimensions, is illustrated in Fig. 2.2.

All recently advanced theories of muscular contraction agree that in all likelihood the actual contractile process takes place at the junctions

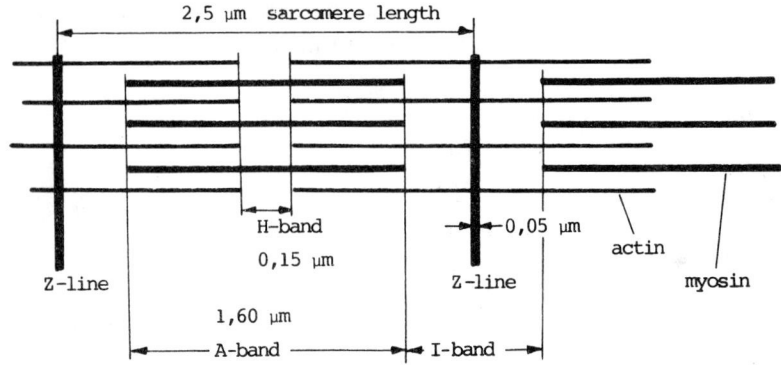

FIG. 2.2: Longitudinal section through fibril showing the molecular substructure (adapted from Huxley, 1969). The cross-bridges connecting the filaments are not shown

between myosin and actin (see Chapter 1). Having localized the points of force production in the muscle it is now possible to proceed to the identification of the mechanical structure elements to be used in the model.

If the muscle is not stimulated neurally, its tension due to the contractile process is extremely small (D.K. Hill, 1968). Thus practically all the tension observed when stretching the resting muscle will be due to elastic structures which lie parallel to the force-producing sarcomeres. These structures are: the sarcolemma of the individual fibre and all outer connective tissue sheaths (endomysia, perimysia, epimysia and fasciae). Banus & Zetlin (1938) have shown that the fasciae are the most important contributors. We are thus presented with a set of *parallel elastic elements* (PEs), some of which are to be considered in parallel with each sarcomere of each fibre and others only in parallel with the whole muscle. Because these tissues move in fluid, it will also be necessary to include appropriate damping elements in the model.

When stimulation takes place the contractile machinery produces tension which, via elastic structures, is transferred to the endpoints of the muscle. These structures are termed *series elastic elements* (SEs). Obvious anatomical sites for these elements are the tendons at the ends of the muscle. These tendons do, in fact, account for the major part, but by no means all, of the series elasticity. Jewell & Wilkie (1958) have clearly demonstrated that an appreciable amount of stretching occurs outside the tendinous regions, along the muscle fibres. The actual locations of these additional series compliances is still a matter of controversy. Jewell & Wilkie and Szent-Györgyi (1953) suggest the Z-discs as the most probable sites. In a more recent paper, Huxley & Simmons (1971) propose that two elastic elements (one of which is damped) in series are located within that part of the myosin molecule which projects out of the myosin backbone. This would appear to contradict the findings of Jewell & Wilkie that the series elastic element does not change substantially with either the muscle length (i.e. with the degree of overlap of the filaments), the temperature or the time after a single shock. However, recent findings by Flitney and Hirst (1978), and Morgan (1977) provide conclusive evidence that part of the series elastic structure is indeed located within the cross-bridges, and that this part depends on the active state of the muscle and on the degree of filamentary overlap (Flitney and Hirst, 1978, Fig. 8).

The following picture of the series elastic elements thus emerges: at both ends of each fibre are lightly damped SEs representing the tendinious parts of the fibre, and two lightly damped SEs, representing respectively the elastic structures within the cross-bridges and the Z-line of the sarcomere, must be added to the sarcomere model of each fibre.

The remaining structure of the sarcomere may now be regarded as constituting a pure *contractile element* (CE).

Finally, the mass distribution along the sarcomere has to be established. It is obvious from Fig. 2.2 that all actin filaments attached to a Z-disc must move with the same velocity as the disc, relative to some reference frame. This can, however, not be asserted with any certainty of the fluid filling the space between the protein filaments. This is not a serious drawback since it can be shown that, within certain limits, the choice of the value of the sarcomere mass is not critical or even significant as regards the model responses. It appears, therefore, permissible to locate the total sarcomere mass at the Z-disc. Also, one does not have to be overly concerned about the mass distributed along the fibres which is due to connective tissue; for the following reason.

It is known (see Chapter 1) that the contracting muscle obeys the constant-volume relation. Let the fibre at a given moment have length $x$ (rest length $\bar{x}$) and assume that its form is approximately cylindrical with radius $r$ (rest value $\bar{r}$). The constant-volume relation then implies that

$$r^2 \pi x = \bar{r}^2 \pi \bar{x} = \text{constant},$$

from which it follows that

$$\dot{r} = -\tfrac{1}{2}\bar{r}\dot{x}(\bar{x}/x)^{3/2}/\bar{x}.$$

This means that when the fibre is shortened (or lengthened) near its rest length, as is the case under physiological conditions, the transverse

velocity of the mass particles on the surface of the fibre is approximately given by
$$\dot{r} = -\tfrac{1}{2}\bar{r}\dot{\bar{x}}/\bar{x}.$$

Now, fibre rest lengths vary from 5 mm (multifidus muscle in man) to about 400 mm (sartorius muscle) while the fibre rest diameters are typically in the region of 100 microns (Starling & Evans, 1968). Thus, for a fibre of average length,
$$|\dot{r}| < 0.001|\dot{x}|,$$
i.e. any influence of transversely moving mass particles may be neglected. This completes the identification of structural elements of the muscle. These elements may now be assembled into interconnecting functional blocks which make up the mechanical equivalent of the skeletal muscle. The result is shown in Fig. 2.3.

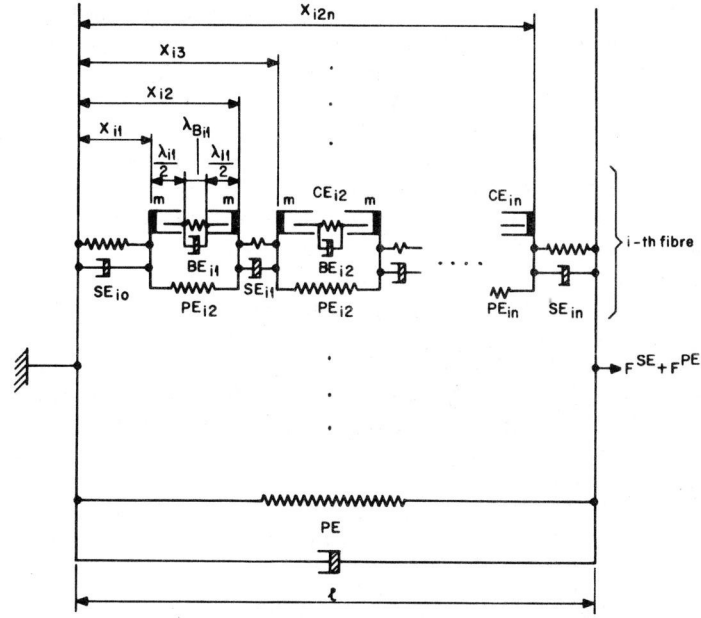

**FIG. 2.3:** Mechanical simulation model of skeletal muscle with the model of the i-th fibre shown in detail. The symbols $SE_{ik}$, $BE_{ik}$, $CE_{ik}$, and $PE_{ik}$ denote respectively the series elasticities due to the Z-discs and cross-bridges, the contractile element, and the parallel elastic element of the k-th sarcomere within the i-th fibre. $SE_{io}$ and $SE_{in}$ denote the series elastic elements due to the tendinous fibre parts, while PE is the parallel elastic element modelling the fibre-external structures. The position coordinate of the ij-th mass element m of the ik-th half-sarcomere is labelled $x_{ij}$, and $l$ is the total length of the muscle

The model displayed in Fig. 2.3 requires some remarks. Firstly, it can be seen that the parallel elastic elements $PE_{ik}$, $k=1,\ldots,n$, do not contain damping components. This is because the sarcolemma, which is mainly responsible for these elements, is attached to each Z-disc and hence does not permit appreciable movement between the membrane and the outer filament layers. Secondly, it can be seen that the sarcomere masses have been split into two parts, located to the left and right of the Z-discs, respectively. The reason for this arrangement is that the Z-discs have some series elasticity, as discussed earlier. Thirdly, it has been tacitly assumed that all fibres in the model are identical with respect to their components, an assumption that is justified only to a limited degree but is essential in order to reduce the complexity of the model, which otherwise could become prohibitive.

The next step in defining the model structure is to establish the functional properties of all the elements involved. This will be done in the next chapter.

# CHAPTER 3

# Functional characteristics and models of the muscle elements

In the preceding chapter three basic elements have been identified which make up the structure of the muscle: the parallel elastic element (PE), the series elastic element (SE) and the contractile element (CE). Although categorically they do not belong to this classification, the mass elements must also be mentioned here as an integral part of the muscle structure. The functional properties of these three basic structural elements will now be dealt with analytically.

## 3.1 FUNCTIONAL CHARACTERISTICS AND MODELS OF THE PARALLEL ELASTIC ELEMENTS

We shall treat the parallel elastic element (PE) of the total muscle first. The PE is characterized by its force-extension relationship and by the function which expresses its dependence on the velocity of the length change. Since the damping element is assumed to be a component in parallel to the elastic one, it is possible to write for the force across the combined elements:

$$F^{PE} = f_1(\zeta) + f_2(\zeta,\dot{\zeta}), \qquad (3.1)$$

where $\zeta = (\ell-\ell_o)/\ell_o$ denotes the relative elongation of the total element, and $\ell_o$ designates the rest length of the PE.

Extensive experiments on the tensile properties of resting skeletal muscle have been carried out by Yamada (1970). The stress-strain (load-extension) curves have been obtained for a variety of human and animal muscles devoid of their fasciae. These curves all have the same characteristic shape, although the elongations corresponding to a given stress value vary over a wide range for different muscles. A typical stress-strain curve is shown in Fig. 3.1.

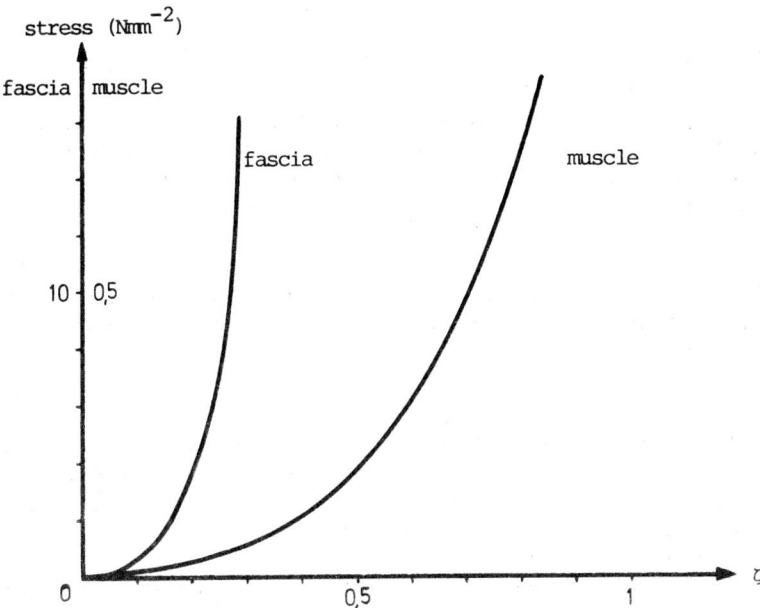

**FIG. 3.1:** Typical stress-strain curves for skeletal muscle and fascia

Yamada has also investigated the functional behaviour of human and animal fasciae (see Fig. 3.1). It was found that the fasciae are much less compliant than muscle tissue (i.e. connective tissue sheaths inside the muscle) and it would thus appear that in the muscle *in situ* the fascia is the distinctly dominating structure in the parallel elastic element.

The curves obtained by Yamada are very similar to those obtained by other investigators. Jewell & Wilkie (1958), for instance, worked on the frog sartorius muscle and for the parallel elastic element found a characteristic curve that resembles that of human muscle very closely. Further published experimental evidence for this type of nonlinear behaviour of the elastic component of the PE is due to Hefner & Bowen (1967).

The question now arises as to which mathematical relationship should be used to describe the nonlinear behaviour of the elastic component of the PE (i.e. an appropriate function $f_1(\zeta)$ is required). Ideally, this function should be derived from basic properties of the biological material involved. A number of attempts have recently been made to relate the stress-strain-history relations of soft tissue to the morphological properties of the underlying structure. A constitutive equation for collagen fibres (which are present in considerable numbers in connective tissue) has been proposed by Haut & Little (1972). The authors used the 'quasi-linear' viscoelastic law suggested by Fung (1968) to derive a set of equations which are then used to test the response of specific biological tissues (rat tails). The predictions of the model were in reasonable agreement with the experimental results except for sinusoidal loadings of the specimen. It should be noted that this treatment includes the time-independent elastic response (the function $f_1(\zeta)$ in the present context) as well as a time-dependent relaxation function. A similar total response will be expected from the function $F^{PE}(\zeta,\dot{\zeta})$.

An attempt to find a general potential function, which would describe the response of isotropic biological material to large stresses, has been made by Snyder (1972b). The author sets out by using the experimental relation (Fung, 1968)

$$dT/d\lambda = aT + k, \qquad (3.2)$$

where T is the Lagrangian stress and $\lambda = \zeta+1$. Snyder then proposes a more general form and uses it to study the response of a muscle fibre to torsional loading. His analysis may, however, be criticized on the grounds that 'it is almost axiomatic that biological tissues are *not* isotropic' (Fung, 1972).

In 1973, Soong & Huang (1973) published a stochastic model for biological tissue elasticity in simple elongation. Their analysis is based on the fact that the major force-bearing components of biological tissues are collagen and elastin, and that these are apparently randomly arranged in biological structures. Their mathematical model (which is an 'elastic response model' only) yields reasonable predictions of experimental data but is still too complicated to be used as a model in a more complex set-up.

The writer (Hatze, 1974) derived the expression

$$k_1(\exp(c_1\zeta)-1)$$

for the function $f_1(\zeta)$, where $k_1$ and $c_1$ are constants. This function is the result of a least-squares fitting procedure applied to a series of experimentally obtained data. Now, it is easily seen that the above expression is, in fact, a solution of (3.2) with the appropriate initial condition (at $\lambda = 1$, $T = 0$) and with the appropriate replacement of the constants. Equation (3.2), in turn, is also an experimental relation obtained by plotting the slope of T against $\lambda$, and fitting the result by a straight line (Fung, 1972, p. 186). More recently, Glantz (1974) used the same model to derive a constitutive equation for the passive properties of muscle.

Latest experiments on human muscle *in situ*, performed by the writer, indicate that in general it is not always possible to account for the experimental data by a single exponential term. It has been found that the most general form of the function $f_1(\zeta)$ is

$$f_1(\zeta) = \sum_{i=1}^{n} k_i (\exp(c_i\zeta)-1). \qquad (3.3)$$

This function effectively indicates that the total response observed is the sum of different structural influences, all of which are basically described by the differential equation (3.2).

It remains now to establish the velocity dependence of the parallel elastic element. That such viscous effects do exist in resting muscle has been demonstrated by Alexander & Johnson (1965), Buchthal & Kaiser (1951), Fung (1970), Walker (1960), and others. The usual approach is to regard the damping element as a viscous component (see, for example, Glantz, 1974). Indeed, all experimental evidence (including that gained by the writer) seems to support this view. The function $f_2(\zeta,\dot\zeta)$ thus becomes

$$f_2(\dot\zeta) = k'\dot\zeta, \qquad (3.4)$$

where $k'$ is a constant and $\dot\zeta = (d\ell/dt)/\ell_o$.

The complete modelling equation for the PE is obtained by substituting (3.3) and (3.4) into (3.1):

$$F^{PE}(\zeta,\dot\zeta) = \sum_{i=1}^{n} k_i(\exp(c_i\zeta)-1) + k'\dot\zeta. \qquad (3.5)$$

It may be of interest to contrast the writer's approach with that of Fung (1972). Fung achieves practically the same effect by defining a fairly general time integral function (supposed to be representative of all viscoelastic biological tissue) which yields the tensile stress T on a cylindrical specimen as a function of time. This so-called quasi-linear viscoelasticity function (see above) is derived as follows. The history of the stress response (the 'relaxation function') which is denoted by $K(\lambda,t)$, is factorized into a normalized function of time $G(t)$, with $G(0) = 1$, and the 'elastic response' $T^e(\lambda)$. The relative length $\lambda$ is as defined in (3.2). Thus

$$K(\lambda,t) = G(t)T^e(\lambda).$$

At time $\tau$, the relative length $\lambda$ is $\lambda(\tau)$, so that an infinitesimal change $\delta\lambda(\tau)$ in stretch produces a stress response of magnitude

$$G(t-\tau)(\partial T^e(\lambda(\tau))/\partial\lambda)\delta\lambda(\tau), \quad t > \tau.$$

Summing the contributions of all the past changes until time t, one obtains the tensile stress at time t as

$$T(t) = \int_{-\infty}^{t} G(t-\tau)\, (\partial T^e(\lambda(\tau))/\partial \lambda)\, (d\lambda(\tau)/d\tau) d\tau.$$

While this approach is very suitable for studies on biological materials in which certain conditions are imposed on the specimen (e.g. experiments with sinusoidal strain input defined by $\lambda(\tau) = \lambda_o \sin\omega\tau$), it is not useful if the element in question is part of a more complex system. In such a system the strain input is determined by the instantaneous state of the other elements of the system, which necessitates a modelling equation of the type proposed by the writer.

The characteristics of the sarcomere parallel elastic elements $PE_{ik}$ (Fig. 2.3) will be discussed in Section 3.3.

## 3.2 FUNCTIONAL CHARACTERISTICS AND MODELS OF THE SERIES ELASTIC ELEMENTS

It is comparatively easy to determine the constants for the PE, which appear in (3.5), by studying the relaxed muscle. This no longer applies when dealing with the series elastic elements (SEs). By definition, these elements come into operation only when the muscle contracts. This means that any externally observable changes in the force, length, etc. are due to a complex interplay between the SEs and the contractile elements (CEs), in a region where the influence of the PE is negligible. This greatly complicates the analysis of the SEs. However, these elements have always been of great interest to muscle specialists, and a large number of experiments have been devoted to the elucidation of the characteristics of these components. Basically, the material responsible for the tendinous part of the SEs does not differ greatly from that making up the PE. Elastin and collagen fibres in complex arrangements are again involved, so that the stress response of this part should not differ fundamentally from that of the PE. This has indeed been found to be the case.

Yamada (1970), in his extensive study of the strength of biological materials, subjected human calcaneal tendinous tissue to tests involving increasing stress on the samples. He obtained stress-strain curves that are qualitatively similar to those shown in Fig. 3.1. The only fundamental difference was that the curves for the tendinous tissue had a much greater slope at $\zeta = 0$ and were generally less curved. The same has been observed by Wilkie (1956a), Hill (1950) and Jewell & Wilkie (1958) in studies on the frog sartorius muscle, and by Bahler (1967), Joyce & Rack (1969) and others on mammalian (rat and cat) muscles.

A word may be in order about the methods used to determine the characteristic curve of the elastic component of the combined SE. Basically three methods are in use: the quick-release method described by Wilkie (1956a), Hill's (1950) controlled-release method, and the calculation of points on the stress-strain curve from points on the curve of the isometric tension development. Without going into the details of any of these methods it should perhaps be mentioned that the quick-release method appears to be the most reliable, while the method based on the use of the isometric tension development has been shown to yield values, from the same muscle, which may differ by a factor of two from those obtained by quick-release (McCrorey et al., 1966).

Values of the normalized extensions $\bar{\zeta}$ of the combined SE, measured relative to the muscle's optimum length $\ell$ and at maximum isometric tension (i.e. when the muscle is fully stimulated, at optimum length $\bar{\ell}$ and not changing its length) have been collected by Close (1972) for various muscles. The values are typically in the range 0.06 to 0.07 for the *in vitro* studies of the mammalian muscles investigated. *In situ* studies, however, yielded much smaller values, viz. in the range 0.03 to 0.048. On the other hand, Wilkie (1950) obtained a value of 0.1 on *in situ* measurements of human biceps muscle. In this study, Wilkie used the calculation method, which has been shown to overestimate the extension.

A smaller value of $\bar{\zeta}$ would also be more consistent with the observations made by Yamada (1970) who found that calcaneal tendinous tissue ruptures at extensions exceeding a value of 0.1 (measured relative to the tendon rest length). The corresponding value for isometric extension (at about one third of the rupture stress) would be 0.065. When related to the total muscle length this value corresponds to a $\bar{\zeta}$-value of about 0.03.

The damping component of the SE has been found to be rather small. Bahler (1967) gives a value of about 300 dyne cm$^{-1}$s for the viscous coefficient of rat gracilis anticus muscle, while Woledge (1961) found a range from 200 to 500 dyne·cm$^{-1}$s (i.e. 0.2 to 0.5 Nm$^{-1}$s) for the frog sartorius muscle.

The equations chosen to model the behaviour of the elastic component of the SE are almost always exponential. Jewell & Wilkie (1958), in their extensive study of the mechanical components in frog's striated muscle, have used an exponential term to describe the first half of the function, and a linear term for the second half. Parmley & Sonnenblick (1967) and more recently Glantz (1974) have also used exponential expressions to characterize this component. Such an approach appears to be justified on the grounds that, as has been mentioned earlier, the tissue making up the tendinous part of the combined SE does not differ fundamentally from that responsible for the PE. In addition, Huxley and Simmons (1971, Fig. 3) have demonstrated that the series elastic component which resides in the cross-bridges also behaves approximately exponentially. Again, the damping component will be assumed to be of a viscous nature. It thus follows that the force across the SE is given by

$$F^{SE} = k_s(\exp(\sigma\delta)-1) + k_s'\dot{\delta}, \qquad (3.6)$$

where $k_s'$, $\sigma$ and $k_s'$ are constants, and the relative elongation $\delta$ is defined by $\delta = (\lambda_s - \lambda_{so})/(\lambda_{sl} - \lambda_{so})$, with $\lambda_s$, $\lambda_{so}$, and $\lambda_{sl}$ denoting respectively the instantaneous length, rest length, and tetanic isometric length of the SE.

Relation (3.6) may be rewritten in a more practical form. Let $\bar{F}$ denote the maximum isometric force produced by the contractile element CE which lies in series with the SE under consideration. Then $k_s = \bar{F}/(\exp(\sigma)-1)$, and (3.6) becomes

$$F^{SE} = \bar{F}(\exp(\sigma\delta)-1)/(\exp(\sigma)-1) + k_s'\dot{\delta}. \qquad (3.7)$$

Relation (3.7) describes the nonlinear viscoelastic behaviour of a typical series elastic component in terms of the constants $\sigma$, $k_s'$, and $\bar{F}$, the latter characterizing a property of the associated contractile component.

As indicated above (see also Huxley and Simmons, 1971), a similar expression may be used to describe the viscoelastic properties of the series elastic structures $BE_{ik}$ (Fig. 2.3) residing in the cross-bridges. The force across *one such cross-bridge series elastic element* is thus given by

$$f^{BE} = \bar{f}(\exp(\sigma\delta)-1)/(\exp(\sigma)-1) + c_s'\dot{\delta}, \qquad (3.8)$$

where $\bar{f}$ is the isometric force produced by that cross-bridge, $\delta = (\lambda_B - \lambda_{Bo})/(\lambda_{B1} - \lambda_{Bo})$, with $\lambda_B$, $\lambda_{Bo}$, and $\lambda_{B1}$ denoting respectively the instantaneous length, rest length, and tetanic length of the cross-bridge SE, and $\sigma$ and $c_s'$ are constants. For reasons that will become apparent presently, it will not be necessary here to decompose the elements $BE_{ik}$ into a damped and an undamped component, as Huxley and Simmons (1971) have done.

We shall deal with expressions (3.7) and (3.8) again in the next chapter, where the constants $\lambda_{so}$, $\lambda_{s1}$, $\lambda_{Bo}$ and $\lambda_{B1}$ will be expressed in terms of a set of experimentally observable parameters.

## 3.3 FUNCTIONAL CHARACTERISTICS AND MODEL OF THE CONTRACTILE ELEMENT

For the purpose of the present discussion it will be assumed that all of the contractile elements $CE_{ik}$, $k=1,\ldots,n$, displayed in Fig. 2.3 are identical. Hence 'contractile element CE' shall mean a typical contractile element $CE_{ik}$ of a muscle fibre sarcomere, having length $\lambda_{ik}$ (see Fig. 2.3) and mass 2m. Note that the cross-bridge series elastic element $BE_{ik}$, which connects the two half-sarcomeres of $CE_{ik}$, is *not* part of the $CE_{ik}$.

The contractile element is the only active component in the model. Its behaviour is extremely complex and depends nonlinearly on its length, contractive history, the rate of length change, the degree of stimulation, and its temperature (in what follows it will always be assumed that the temperature is 37°C). It is, however, possible to reduce the description of the behaviour of the contractile element (CE) to the discussion of three basic functions: the active-state function, the filamentary-overlap function, and the velocity-dependence function. This will now be done.

It is now generally accepted that the force produced by the CE is due to the interaction, via so-called cross-bridges, of the myosin and actin filaments of the muscle fibre. In fact, the cross-bridges are part of the myosin filament, being the heavy-meromyosin (HMM) subunits projecting out of the light-meromyosin assemblage which constitutes the backbone of the myosin filament. The HMM subunits, in turn, each consist of two rod-like subunits HMM-S2, each of which carries a globular head subunit HMM-S1 (Lowey, Slayter, Weeds and Baker, 1969). The globular heads to which the energy-providing ATP molecules are presumed to bind in the presence of Ca ions, are thought to provide the direct link with the actin filaments.

The contractive force produced by a muscle fibre is thus equal to the sum of the forces produced by all the cross-bridges in one half-sarcomere of the fibre, at any instant of time. Because the propagation velocity of the Ca ions moving from the terminal cisternae into the sarcoplasm is finite (Jöbsis and O'Connor, 1966), the onset of the contractive cycle of different sets of cross-bridges along the myosin filament upon stimulation will be successive. That this is indeed the case has been demonstrated by Huxley and Taylor (1958).

Instead of considering the contractive state of each set of cross-bridges, it is possible to define an 'average contractive force' of all the cross-bridges at a given instant of time (Hatze, 1973). The total fibre force is then equal to the sum of all the cross-links existing in one half-sarcomere, each of which cross-links produces that average contractive force.

The number of cross-links formed is a function of the active state of the fibre (defined below), the degree of filamentary overlap (Gordon, Huxley and Julian, 1966) and, presumably, the velocity of shortening or lengthening of the contractile element (A.F. Huxley, 1957). On the other hand, the average force output of the cross-bridges is postulated to depend on the velocity of the interfilamentary movement and on certain intermolecular forces.

These intermolecular forces exhibit a cyclic behaviour in accordance with the attachment-detachment-reattachment cycle of the cross-bridges (Huxley, 1957; Hatze, 1973). Owing to the asynchronous actions of the various cross-links, these individual force fluctuations do not appear externally, so that the average force output of an activated cross-bridge becomes a function of the velocity of the interfilamentary movement only.

In thus follows that the force output $f^{CP}$ of the contractile proteins of a typical muscle fibre is equal to the average force output $G(\dot{\lambda})$ of a typical cross-bridge, multiplied by the number $N\Omega$ ($0 < \Omega \leq 1$) of cross-links active in a half-sarcomere, at any instant of time, i.e.

$$f^{CP} = N\Omega G(\dot{\lambda}), \qquad (3.9)$$

where N denotes the total number of cross-bridges present in a half-sarcomere, $\dot{\lambda} \triangleq d\lambda/dt$, and $\lambda$ is the instantaneous length of a sarcomere as shown in Fig. 2.3. Note that $\Omega$ designates the fraction of active cross-bridges.

As indicated above, the relative number $\Omega$ of active cross-links is a function of the active state, the degree of filamentary overlap, and the velocity $\dot{\lambda}$ of shortening or lengthening of the contractile element (CE). Let the normalized length $\xi$ of the CE be given by

$$\xi = \lambda/\bar{\lambda}, \qquad (3.10)$$

where $\bar{\lambda}$ is that sarcomere length at which the force output of the fully stimulated and isometrically contracting CE attains a maximum. Then we may define a normalized function $k(\xi)$, $0 < k(\xi) \leq 1$, which not only reflects the degree of filamentary overlap but also the influence on the number of active cross-links of the distortions caused by the collision between Z-lines and myosin filaments at very short sarcomere lengths, and the overlap of actin filaments at intermediate sarcomere lengths (Gordon et al., 1966). The filamentary-overlap function $k(\xi)$ will be discussed in detail in Section 3.3.2.

The second factor influencing the number of active cross-bridges is the active state, denoted by q. Again, $q(\cdot)$ is a normalized function of several variables, and will be discussed in Section 3.3.1.

Finally, the velocity-dependence of $\Omega$ will be expressed by the normalized (and as yet unspecified) function $H(\dot{\lambda})$.

Since the three factor functions determining $\Omega$ each contribute individually to the total result, we have that

$$\Omega = q(\cdot)k(\xi)H(\dot{\lambda}), \qquad (3.11)$$

so that (3.9) can be written as

$$f^{CP} = q(\cdot)k(\xi)g(\dot\lambda) ,\qquad(3.12)$$

where we have defined the non-normalized combined velocity-dependence function $g(\dot\lambda)$ by

$$g(\dot\lambda) \triangleq NH(\dot\lambda)G(\dot\lambda).\qquad(3.13)$$

This function will be discussed in detail in Section 3.3.2.

Equation (3.12) is an expression for the total force output of the *contractile proteins* of a fibre sarcomere, for all contractive modes. However, the force output $f^{CP} \triangleq f^{CE}$ of the contractile element is not identical with the force output $f^C$ as *observed externally*, for the following reason.

Gordon *et al*. (1966) and Ramsey and Street (1940) have convincingly demonstrated the existence of an internal force $f_o(\xi)$ 'tending to extend the fibre and therefore presumably subtracting from whatever tension is developed by the filaments during contraction' (Gordon *et al*. 1966, p. 187). These authors have also suggested that this internal force may be attributed to the deformation of the sarcolemma accompanying the increase of the fibre diameter in shortening. From the known structural properties of the sarcolemma one can infer that the force $f_o(\xi)$ will have an exponential characteristic, i.e.

$$f_o(\xi) = a_5 - a_5(\exp\{a_6'(\xi-1)\}-1) = a_5(2 - \exp\{a_6'(\xi-1)\})\qquad(3.14)$$

$a_5$ and $a_6'$ being constants. Clearly, the force output $f^{CE}$ of the contractile machinery will be reduced by this internal resistance, and the *externally observable* contractile force $f^C$ is therefore given by

$$f^C = q(\cdot)k(\xi)g(\dot\lambda) - f_o(\xi).\qquad(3.15)$$

From Fig. 2.3 we see that $f_o(\xi)$ is due to a typical parallel elastic element $PE_{ik}$ which has been attributed to the sarcolemma of the fibre (see Chapter 2). Note that $f_o$ is actually a function of both $\xi$ and $\lambda_B$ (see Fig. 2.3), but that we shall be able to represent $\lambda_B$ by a constant, as will be demonstrated later.

We now have an expression (Equation (3.15)) for the total force output $f^C$ of the contractile structures of a muscle fibre. However, the functions $q(\cdot)$, $k(\xi)$, and $g(\dot\lambda)$ are still unspecified. In the next two sections we shall give a detailed derivation of the expressions which define these functions.

### 3.3.1 The excitation dynamics

The excitation dynamics comprises the dynamics of the interfilamentary

Ca-ion concentration and its influence on the active state of the muscle fibre. We shall begin by discussing the active state.

The active-state function was originally defined by Hill (1938). This definition was not exact, and was given in verbal terms only. Hill pictured the active state as some operational ability of the muscle which abruptly appeared when the muscle was stimulated, and more slowly disappeared on cessation of the stimulus. In a later paper (Hill, 1949), he defined the intensity of the active state as the tension which the CE could develop, without lengthening or shortening, at any point of time after the beginning of the excitation. Although this definition is more precise than the original one and excludes the influence of the SE and the force-velocity function, it still does not specifically exclude the influence of the length-tension relation. Hill also proposed two other characteristics of the active state: the intensity of the active state is reached immediately upon stimulation (i.e. in zero time) and its magnitude is maximal after the first stimulus. Both these assumptions have been disproved (see below) since Hill's paper was published.

Several methods have been developed to determine the time course of the active state as defined by Hill. These methods have been reviewed by Bahler et al. (1967) and Close (1972), and will not again be described here.

In order to derive the active-state function of the present model it is necessary to briefly describe the events leading to the onset of the contraction. The sequence starts with the arrival of a nerve signal at the motor end plate of the fibre and the subsequent propagation of the fibre action potential along the fibre surface and down the transverse tubular system (Gonzalez-Serratos, 1966). The transverse tubular system (T-system), which is located within the Z-discs, converts the action potential into a depolarization signal which acts across the walls of the tubular network (Adrian, Chandler & Hodgkin, 1969; Falk, 1968). Lumped and distributed electrical networks consisting of capacitors and resistors have been used by Falk & Fatt (1964) and Schneider (1970) to simulate the T-system (for a review of these modelling attempts see Peachey, 1973). Next, the depolarization of the T-membranes causes the release of calcium ions ($Ca^{2+}$) from the sarcoplasmic reticulum, a membrane structure situated within the fibre. (The whole question of the active transport of calcium ions in sarcoplasmic membranes has been reviewed by Inesi, 1972.) On penetration of the interfilamentary space by the calcium ions, ATP (adenosinetriphosphate) hydrolysis is initiated and, simultaneously, $Ca^{2+}$ binds to the calcium-binding subunit of the troponin molecule. This causes neutralization of the inhibitory effect of the TN-I subunit, enabling the myosin head to bind to an actin monomer (Ebashi & Kodama, 1965; Hartshorne & Dreizen, 1973; Perry, Cole,

Head & Wilson, 1973; and others). It must be stressed that this is an extremely simplified representation of the excitation-contraction coupling process, and can only be justified on the grounds that the present study is not concerned with the physiological details of the process under consideration but with the mathematical modelling of the events taking place.

In accordance with Ebashi and Endo (1968, p. 139) we *define the active state* q *to be the relative amount of Ca bound to troponin*. If the maximum number of potential interactive sites on the actin filament is exposed by the action of calcium, then q = 1, while in resting muscle q = $q_o$. From the above definition of the active state it is clear that the isometric tension developed by a muscle fibre at a given length λ of the CE is directly proportional to q.

Define γ to be the difference between the real free Ca ion concentration $γ_r$ and the free Ca ion concentration $γ_o$ in the resting fibre, i.e. γ = $γ_r$ − $γ_o$. Since the value of $γ_o$ is about $1 \times 10^{-9}$ to $8 \times 10^{-9}$M (Ebashi and Endo, 1968; Gillis, 1969) and the value for the mechanical threshold is in the region of $4 \times 10^{-7}$ to $8 \times 10^{-7}$M (Ebashi and Endo, 1968), it is clear that for all practical purposes we have γ = $γ_r$. Let now p = dq/dγ denote the Ca concentration rate of change of the active state q. For the process of the binding of the Ca ions to the troponin-tropomyosin complex in the presence of a varying Ca concentration, one would expect, at a fixed length γ, an increment δp for a given increment δγ of the free Ca ion concentration to be proportional to the difference between the maximum and the present value of q, controlled by a negative feedback loop. There is strong experimental evidence (Bahler *et al.*, 1967; Jewell and Wilkie, 1960; Rack and Westbury, 1969) that such a relationship may also depend on the length λ of the CE. The above state of affairs may be expressed by

$$δp/δγ = r_1(ξ)(1-q) - r_2(ξ)p , \qquad (3.16)$$

where $r_1(ξ)$ and $r_2(ξ)$ are functions (yet to be determined) of the normalized length ξ, defined by (3.10).

A relationship of the form (3.16) has indeed been found repeatedly in experiments on the isometric tension development of skinned fibres when the Ca concentration of the bathing solution was varied (Fig. 3 of Ebashi and Endo, 1968; Julian, 1971).

Letting δγ→ 0 in (3.16), and holding ξ constant, we obtain the differential system

$$\begin{aligned} dq/dγ &= p, & p(0) &= 0, \\ dp/dγ &= r_1(ξ)(1-q) - r_2(ξ)p, & q(0) &= q_o, \end{aligned} \qquad (3.17)$$

where $q_o$ denotes the resting active state of the fibre with a value of about 0.005.

Putting $r_1(\xi) = \rho_1^2(\xi)$, $r_2(\xi) = 2\rho_2\rho_1(\xi)$, $\rho_2 > 1$, and introducing the variable transformation

$$\omega \triangleq \rho_1(\xi)\gamma, \qquad \rho_1(\xi) \neq 0, \qquad (3.18)$$

it is easily shown that the analytical solution of the system (3.17) is given by

$$q(\omega) = 1 - (1-q_o)[m_1\exp(m_2\omega) - m_2\exp(m_1\omega)]/(m_1-m_2), \quad (3.19)$$

where

$$m_{1,2} = -\rho_2 \pm (\rho_2^2 - 1)^{1/2}, \qquad \rho_2 > 1.$$

On reintroducing $\gamma$, equation (3.19) becomes

$$q(\xi,\gamma) = 1 - (1-q_o)[m_1\exp\{m_2\rho_1(\xi)\gamma\} - m_2\exp\{m_1\rho_1(\xi)\gamma\}]/(m_1-m_2). \quad (3.20)$$

From (3.18) it is clear that the $\xi$-dependence of $q$ is, in fact, introduced by the pseudo Ca ion concentration $\omega(\xi,\gamma)$. However, the factor model (3.18) is only *one* possible means of expressing the dependence on $\xi$ of the Ca concentration, albeit one that will be found useful later. Hence, in general, we must express the Ca response $\gamma$ as $\gamma = \gamma(\xi,t)$, of which the model $\gamma(\xi,t) = \rho_1(\xi)\gamma(t)/r_3$ constitutes a special case ($r_3$ is a constant).

There is strong experimental evidence that the release of Ca ions from the sarcoplasmic reticulum and hence the free Ca concentration $\gamma$ in the interfilamentary space is the response, to a spike-like stimulus, of a second-order system. Indeed, Jöbsis and O'Connor (1966) have directly measured the time course of $\gamma$ in the sartorius muscle of the toad (see their Figs. 1-3) and obtained a function which closely resembles the delayed response of an overcritically damped second-order system. The investigations of Hodgkin and Horowicz (1960) also point in this direction. Finally, Ebashi and Endo (1968) point out that the depolarization of the tubular membrane seems to have two effects on the sarcoplasmic reticulum: one causing a release of Ca, the other a re-uptake of Ca.

More specifically, experimental evidence has now accumulated (Constantin and Podolsky, 1966; Natori, 1965; Peachey, 1965) strongly suggesting, as the direct cause of Ca release from the sarcoplasmic reticulum, a direct depolarization of the membrane of the sarcoplasmic reticulum by the depolarizing potential of the T-tubular system. Estimates of the size of the sarcoplasmic membrane potential range from 0.5-1.5 mV (Ebashi and Endo, 1968).

We are therefore justified in regarding $\gamma$ as the output of an electrical second-order system, the input of which is the depolarizing potential $V_T\beta(t)$ of the T-system. Hence $\gamma(\xi,t)$ is defined by

$$\ddot{\gamma} + (c_1\dot{\gamma} + c_2\gamma)/\rho^*(\xi) = c_3 V_T\beta(t), \quad \gamma(0) = \dot{\gamma}(0) = 0, \quad (3.21)$$

where $c_1, c_2, c_3$ are constants, $V_T$ is about 0.05 V (using an estimate of Ebashi and Endo, 1968), and the normalized density function $\rho^*(\xi)$ is given by (3.26) below.

As regards the signal $V_T\beta(t)$ in the interior of the T-system, it is now fairly certain (Huxley and Peachey, 1964) that this is due either to direct conduction of the action potential of the surface membrane, or to electronic spread of the surface potential down the T-system. The form of the potential $V_T\beta(t)$ will therefore resemble that of the surface action potential. The latter, in turn, is known to be the result of the electrochemical transmission of the nerve impulse $V_N\alpha(t)$ arriving at the motor endplate of the fibre (see, for example, Del Castillo and Katz, 1956). The work of Adrian, Chandler and Hodgkin (1969), Falk (1968), and Falk and Fatt (1964) indicates that the neuromuscular junction and the T-system are to be regarded as complex electrical networks. However, for the purose of obtaining the response $V_T\beta(t)$ to the nerve input $V_N\alpha(t)$, the total membrane system may (as will be shown below) be modelled as a lumped second-order system described by

$$\ddot{\beta} + c_4\dot{\beta} + c_5\beta = c_6 V_N\alpha(t), \quad \beta(0) = \dot{\beta}(0) = 0, \quad (3.22)$$

where $c_4, c_5, c_6$ are constants and $V_N = 90$mV (Eccles, 1973).

Although the nerve potential $V_N\alpha(t)$ must also be regarded as the response of an electrical second-order system, its shape may be closely approximated by a half-sine wave with a half-period of 1 ms (see, for example, Fig. 4.24 of Walsh, 1964). Normally, a muscle fibre is stimulated by trains of nerve impulses occuring at variable intervals $\tau$, i.e.

$$\alpha(t) = \sin 1000\pi(t-t_i) \quad \text{for} \quad t_i \leq t \leq t_i + 0.001,$$
$$= 0 \quad \text{otherwise}, \quad i=1,2,\ldots, \quad (3.23)$$

where $\tau = t_{i+1} - t_i$, and $\tau^{-1}$ is the stimulation rate.

In general, there are time delays between the onsets of $\alpha(t)$, $\beta(t)$, $\gamma(t)$ and $q(t)$, owing to finite conduction velocities in the membrane system and limitations in the rates of the chemical reactions involved. We are, however, here concerned only with the effects which the train of stimulating nerve impulses has on the active state of the muscle fibre, i.e. we regard the onset of the active state as the initial event which is elicited by, but occurs simultaneously with, the onset of the first stimulating nerve volley.

We still have to obtain an appropriate expression for $\rho^*(\xi)$ (see eq.(3.21)). It is well known that as the fibre shortens from a given length, its cross-sectional area increases owing to the constant-volume relation (Dragomir, 1970). If it is assumed that at a given overall concentration the amount of free Ca ions present in the filamentary-overlap region remains approximately constant, then the *density* $\rho_1(\xi)$, relating to a unit of volume of interfilamentary space, should increase with increasing $\xi$ in accordance with the changes occurring in the geometry of the filament lattice and the degree of filamentary overlap. This implies that

$$\rho_1(\xi) = k_1/V(\xi) = 4k_1/d^2\pi\bar{\lambda}(k_2 - \xi) \triangleq r_3\rho^*(\xi), \qquad (3.24)$$

where d is the fibre diameter, $\bar{\lambda}(k_2 - \xi)$ the filamentary overlap, $k_1$, $k_2$, and $r_3$ are constants, and $\rho^*(\xi)$ denotes the *normalized Ca density*. The constant-volume relation yields

$$\lambda d^2\pi/4\bar{\lambda} = \xi d^2\pi/4 = \text{constant}, \qquad (3.25)$$

which may be substituted into (3.24), eliminating d in favour of $\xi$. To allow for the possibility of the occurrence of nonlinear effects (such as the collision of the myosin filaments with the Z-lines at short lengths) we include the exponential constant s, which in the linear case is equal to unity. The resulting relationship is then the required expression for $\rho_1(\xi)$, i.e.

$$\rho_1(\xi) = r_3\rho^*(\xi), \quad \rho^*(\xi) = [\check{\xi}^s - 1]/[(\check{\xi}/\xi)^s - 1], \quad 0.58 \leq \xi \leq 1.8, \qquad (3.26)$$

where all the constants appearing in (3.24) and (3.25) have been absorbed into the constants $r_3$ and $\check{\xi}$, and $\rho^*(\xi)$ (needed in (3.21)) is normalized such that $\rho^*(1) = 1$.

It must, of course, be realized that the postulate (3.24) and hence the relation (3.26) may be a gross over-simplification of a much more complicated process. In fact, the length dependence of the active state may have causes entirely different from those assumed here. However, as will be shown later, the predictions of the model are in good agreement with experimental results, so that there is no need at this stage to make any other assumptions regarding the length dependence of $\gamma$.

The question now arises as to which values to assign to the constants appearing in Equations (3.21), (3.22) and (3.26). These values can be obtained most efficiently by simulating the system (3.21–3.23) on an analog computer and substituting the resulting function $\gamma(\xi,t)=\omega$ into Equation (3.19). Since, however, all of these constants are specific for a given type of muscle working at a certain temperature, it was decided to perform the simulation for fast and slow human muscle fibres and for a temperature of 37°C.

Figure 3.2 depicts the simulation results for fast muscle fibres. It can be seen that the peak of the action potential $V_T\beta(t)$ (second curve from the left in both photographs) occurs at about 1 ms and that the response declines to half its maximum value at 2.6 ms. This is the form of the action potential as it is usually found in mammalian muscle (see, for example, Fatt, 1959). The time course $\gamma(\xi,t)$ of the free Ca concentration (Fig. 3.2), together with the active state, will be discussed in detail in chapter 6, for the whole range of permissible values of $\xi$. The present simulations were performed for a fixed value of $\xi = 1$.

**TABLE 3.1: Values of muscle-specific constants appearing in the differential equations**

| Fibre type | $c_1$ | $c_2$ | $c_3$ | $c_4$ | $c_5$ | $c_6$ | $r_3$ |
|---|---|---|---|---|---|---|---|
| fast | $2.24\times 10^4$ | $2.50\times 10^5$ | $8.6\times 10^{10}$ | $1.84\times 10^4$ | $7.36\times 10^6$ | 10.333 | $1.14\times 10^5$ |
| slow | $2.24\times 10^4$ | $0.71\times 10^5$ | $9.2\times 10^{10}$ | | | | |

| $\rho_2$ | $\tilde{\xi}$ | s | $q_o$ | c | m |
|---|---|---|---|---|---|
| 1.05 | 2.90 | 1 | 0.005 | $1.373\times 10^{-4}$ | 11.25 / 3.67 |

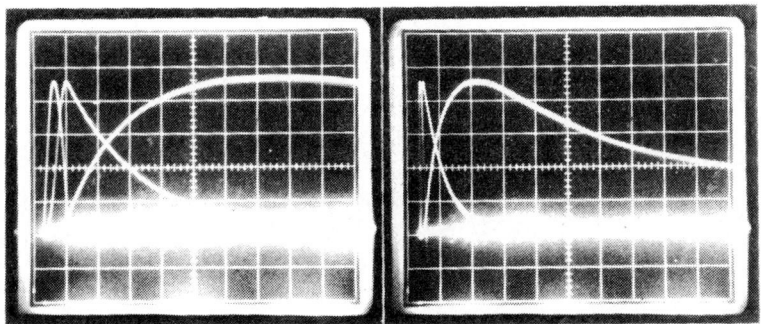

**FIG. 3.2:** Simulation responses $V_T\beta(t)$ and $\gamma(t)$ for fast muscle fibres. Left photograph: First curve (peak furthest left) is the nerve input $V_N\alpha(t)$, the second curve (second peak from left) is the action potential $V_T\beta(t)$, while the third curve (broad peak on the right) depicts the free Ca ion concentration $\gamma(t)$. All curves start at the beginning of $V_N\alpha(t)$. Ordinate: arbitrary units; abscissa: 1.2 ms/div. Right photograph: same responses as on left photograph with the time scale suppressed by a factor of 4 (from Hatze, 1977a)

The values of the constants obtained from the analog simulation for fast and slow human muscle fibres have been collected in Table 3.1.

It is now possible to observe the responses of the system (3.21-3.23) to trains of nerve impulses $V_N\alpha(t)$, delivered at different intervals $\tau$. In other words, we want to observe how the free Ca concentration $\gamma$ in the sarcoplasm varies as a function of time, at selected stimulation rates $\tau^{-1}$. This and all subsequent simulations were performed on a CDC 174 digital computer. The results for fast and slow fibres, and for constant stimulation rates are shown at the top left and right, respectively, of Fig. 3.3. In all cases the value of $\xi$ was equal to unity.

The lower two diagrams of Fig. 3.3 display the active state functions q corresponding to the respective inputs $\gamma(t,\tau^{-1})$.

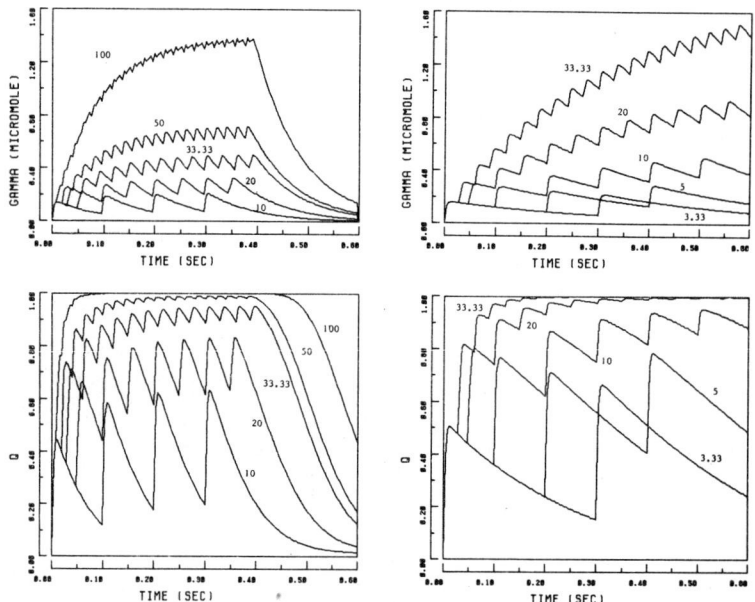

**FIG. 3.3:** Upper part: Free Ca ion concentration $\gamma$ as a function of time for different stimulation rates (indicated by the numbers next to each graph) and for fast (graph on the left) and slow (graph on the right) muscle fibres. The irregularities in the ripple in the top curve of the left-hand graph are due to insufficiently small sampling invertals of the plotter routine and are not genuine. Lower part: Active state function q corresponding to the inputs $\gamma$ as displayed in the upper part of the figure. For the fast fibre, the stimulation ceased at 0.4 s (from Hatze, 1977a)

Figure 3.3 indicates that it may be possible to obtain a 'trend function' for $\gamma(\xi,t,\tau^{-1})$ which is smooth and represents the 'average behaviour' of $\gamma$ in successive time intervals $\Delta t$. Moreover, Figure 3.3 suggests that such a function should approach a maximum value asymptotically, and that both maximum value and initial rate of increase (or decrease, after cessation of stimulation) should be proportional to the stimulation rate. An analysis of the system (3.21-3.23), and of the curves in the upper part of Figure 3.3 reveals that this is indeed the case. In fact, it turns out that for $\xi = 1$ the trend function $\gamma(v,t)$ can be adequately described by the differential equation

$$\dot{\gamma} = m(cv-\gamma), \qquad \gamma(0) = \gamma_o, \qquad (3.27)$$

where m and c are constants given in Table 3.1, and the *relative stimulation rate* v is the first *control parameter* of the muscle fibre, defined by

$$0 \leq v = \hat{\tau}/\tau \leq 1, \qquad (3.28)$$

$\tau^{-1}$ and $\hat{\tau}^{-1}$ denoting the stimulation rate and maximum stimulation rate respectively.

The definition of the maximum stimulation rate $\hat{\tau}^{-1}$ has to be somewhat arbitrary, since for the fast fibre the active state q reaches the value 0.985 at a stimulation rate of 50 c/s (at $\xi = 1$), a value of 0.99992 at 100 c/s, etc. In other words, q(v) approaches unity asymptotically. There is, however, experimental evidence which imposes an approximate upper limit on $\hat{\tau}^{-1}$.

Eccles (1973, p. 51) reports that in mammalian muscle the degree of liberation of acetylcholine transmitter substance per stimulating pulse decreases above stimulation rates of 100 c/s, while 100 c/s is also about the maximum firing frequency observed in fast human muscle (Marsden, Meadows and Merton, 1971). The values for slow muscles are about one-half to one-third of those for fast muscles (Bigland and Lippold, 1954). We therefore define $\hat{\tau}^{-1}$ to have a value of 100 c/s for fast and 28 c/s for slow fibres. These constants should not be regarded as fixed for all muscles but rather as approximate average values.

For the case when $\xi \neq 1$ it can be demonstrated both theoretically and by simulation that the function $\rho^*(\xi)$, as defined by (3.26), multiplies $\dot{\gamma}$, c, and $\gamma$ in (3.27), so that $\gamma(\xi,v,t)=\rho^*(\xi)\gamma(v,t)$. Note that $\rho^*(\xi)$ cancels in (3.27) since it multiplies both sides of the equation. In accordance with (3.26) the input $\rho_1(\xi)\gamma$ to the active state function (3.20) is given by

$$\rho_1(\xi)\gamma = r_3\rho^*(\xi)\gamma(v,t).$$

To summarize, (3.27) describes the *average* calcium dynamics of the CE of a single muscle fibre stimulated at a normalized average rate v. The active state q, in turn, depends nonlinearly on the Ca concentration $\gamma(t)$, as described by (3.20). The dynamics of $\gamma$ and that of q taken together represent the *excitation dynamics* of the contractile element of a single muscle fibre. It must, however, be kept in mind that the active state as defined by (3.20) and (3.27) represents an average property of a set of fibres fired upon asynchronously, as occurs in the total muscle. The exact response to the single nerve volleys stimulating the fibre is, of course, given by the output of the system (3.21-3.23), (3.20), with $\rho_1(\xi) \equiv r_3$ in (3.20), since the length-dependence of $\gamma$ is already included in (3.21).

From Fig. 3.3 it is apparent that

(i) the active state of mammalian muscle does not reach its full intensity after a single stimulus (which is in accordance with the findings of Desmedt and Hainaut, 1968);

(ii) the active state reaches its peak after about 10-12 ms as has been shown experimentally by Bahler *et al.* (1976);

(iii) the decay of the active state q is only approximately exponential in its later part; and

(iv) the relative increase of q as response to a stimulus depends on the present value of q.

It should also be noted that the predicted response of the free Ca ion concentration $\gamma(t)$ to repeated stimulation (upper part of Fig. 3.3) is confirmed by experimental evidence (see Fig. 3 of Jöbsis and O'Connor, 1966). For each value of the relative stimulation rate v, the concentration $\gamma(t)$ approaches a maximum which, by Equation (3.20), implies a maximum value $q_{max}(v,\xi) = Q(v,\xi)$ of the active state. This maximum $Q(v,\xi)$ for the fast muscle fibre has been plotted in Fig. 3.4 as a function of the absolute stimulation rate and for three different parameter values of the fibre length $\ell$.

The graph (Hatze, 1977a, Fig. 4) of the isometric steady-state force output of the fibre, which corresponds to the active state $Q(v,\xi)$ as shown in Fig. 3.4, exhibits the well-known sigmoid shape (Bigland and Lippold, 1954; Milner-Brown, Stein and Yemm, 1973a) as well as the length dependence of this relation, characteristic of mammalian muscles (Rack and Westbury, 1969).

From a purely computational point of view it is more economical to use a simpler approximation of the active-state function $q(\xi,\gamma)$ as defined by (3.20). Such an approximation is given by the function

$$q(\xi,\gamma) = \frac{q_0 + [\rho(\xi)\gamma]^2}{1 + [\rho(\xi)\gamma]^2} \qquad (3.29)$$

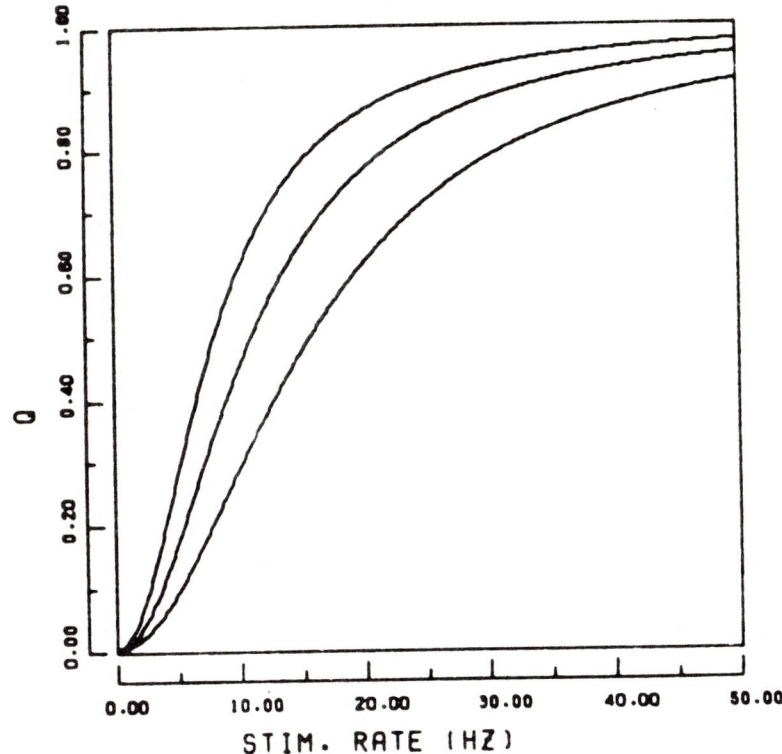

**FIG. 3.4:** Maximum values Q of the active state for the fast fibre as functions of the stimulation rate and for different parameter values of the muscle length $\ell$. The upper curve is for $\ell = 1.08\tilde{\ell}$, the middle curve for $\ell = \tilde{\ell}$, and the lower curve for $\ell = 0.9\tilde{\ell}$

where

$$\rho(\xi) = 66200 \frac{\check{\xi}^s - 1}{(\check{\xi}/\xi)^s - 1} \ . \tag{3.30}$$

In some cases, it has been found that the function

$$q(\xi,\gamma) = \frac{q_o + [\rho(\xi)\gamma]^3}{1 + [\rho(\xi)\gamma]^3} \tag{3.31}$$

with

$$\rho(\xi) = 52700 \frac{\check{\xi}^s - 1}{(\check{\xi}/\xi)^s - 1} \ , \tag{3.32}$$

41

provides a better fit to experimentally determined active-state functions $q(\xi,\gamma)$. For most mammalian muscles values of $\bar{\xi} = 2.90$ and $s = 1$ appear appropriate. If the system (3.21-3.23) is to be simulated, then $\bar{\xi} \equiv 1$ in (3.30) or (3.32), since the length-dependence of $\gamma$ is already accounted for in (3.21).

Hence the *excitation dynamics* of the contractile element of a single muscle fibre is defined by (3.27), i.e. by

$$\dot{\gamma} = m(cv(t)-\gamma), \qquad \gamma(t=0) = \gamma_o,$$

where $v(t)$ denotes the normalized neural control average stimulation rate, and by the corresponding active-state output $q(\xi,\gamma(t))$, as given by (3.29) (or (3.31) in exceptional cases).

### 3.3.2 The contraction dynamics

The contraction dynamics of the contractile element of a muscle fibre is described by the *velocity-dependence function* and by the *filamentary-overlap function*. We shall begin the discussion with the latter.

The influence on the number of active cross-links in the interfilamentary space of the degree of filamentary overlap is one of the most extensively studied features of the contractile machinery and therefore well known. We shall not elaborate on the molecular details of this relationship: the interested reader is referred to the famous paper by Gordon, Huxley and Julian (1966). It is easily verified that, in a purely formal manner, the length-tension relation $k(\xi)$ of the contractile element of an isolated fibre can be adequately described by

$$k(\xi) = 0.32 + 0.71 \exp\{-1.112(\xi-1)\} \times \\ \sin\{3.722(\xi-0.656)\}, \quad 0.58 \leq \xi \leq 1.8. \quad (3.33)$$

This function has a maximum value of approximately unity (at $\bar{\xi} = 1$) and a minimum of zero, and approximates the experimental length-tension curve (Gordon, Huxley and Julian, 1966, Fig. 12) very closely. It must, however, be emphasized that the function (3.33) does *not* necessarily give an adequate description of the length-tension relation of a whole muscle. The reason for this is that the values of $\bar{\xi}$ for the different fibres obey a statistical distribution with sometimes large values of the variance, as has been demonstrated experimentally by Bagust *et al.* (1973). As would be expected for this case, experiment shows (Stevens *et al.*, 1980) that for the whole muscle the function $k(\xi)$ resembles a normal curve and can be adequately modelled by

$$k(\xi) = \exp\left\{-\left[\frac{\xi-1}{s_k}\right]^2\right\}, \qquad (3.34)$$

where again $k(1) = 1$, i.e. the filamentary-overlap function for the whole muscle is also normalized. The parameter $s_k$ determines the 'peakedness' of the curve and its value is specific for a given muscle. A detailed discussion of methods for estimating the value of $s_k$ will follow in Chapter 5. Note that although (3.34) resembles the definition of a normal curve and, indeed, derives its justification from the law of large numbers, it is obviously not a statistical distribution (since the area under the curve is not equal to unity) but a mathematical function representing a property of the whole muscle as contrasted with the single fibre.

The next factor which determines the force output of the fibre is the *velocity of movement* between the actin and the myosin filaments. This is one of the most complicated and least understood characteristics of the muscle. The relation between the velocity of movement and the force production is highly nonlinear and, for hyperfast extensions even discontinuous, as demonstrated by the experiments on isolated fibres by Sugi (1972) and others.

While the whole range of negative velocities (shortening) as well as small to moderately large positive velocities (stretches) can be treated using conventional mathematical techniques, this no longer applies to the region of hyperfast extensions, where the discontinuities occur. There are indications that it may be possible to model this region by applying the methods of catastrophe theory (Thom, 1975). This approach is at present being investigated. In this monograph we shall, however, be concerned only with regions of the velocity-dependence function where no discontinuities occur.

It must be emphasized that the velocity-dependence function $g(\dot{\lambda})$ as defined by Equation (3.13) is *not* identical with the force-velocity relation as it is known in muscle physiology. The term 'force-velocity relation' is actually a misnomer since this relation involves the length $\lambda$ of the CE and its active state q, as will be shown below. The many attempts to model the force-velocity relation are discussed in (Hatze, 1975a). In contrast, the velocity-dependence function depends only on the interfilamentary longitudinal velocity $\dot{\lambda}$ of the CE.

In (Hatze, 1977a) we have demonstrated that the velocity-dependence function $g(\dot{\lambda})$ exhibits a sigmoidal shape and can be described by functions composed of exponentials. Interestingly, the prediction of this model that the curvature of the corresponding force-velocity curve should change even at small negative (shortening) velocities, i.e. before zero velocity is reached, has recently been confirmed experimentally by Edman, Mulieri and Scubon-Mulieri (1976). These authors demonstrate that the force-velocity relation of single muscle fibres, as well as fibre bundles, cannot be fitted satisfactorily by Hill's equation, a fact which invalidates not only Hill's model but most others as well.

There are many functions composed of exponentials and capable of adequately describing the velocity-dependence function $g(\dot{\lambda})$ in its domain of continuity (i.e. excluding hyperfast extensions of the CE). However, the range of these functions is restricted by the requirement that the number of parameters appearing in the defining equation should be minimal. On theoretical grounds the choice of the hyperbolic tangent function is indicated (Hatze, 1977a), but it has been found that more flexibility can be obtained with another function containing the same number of parameters. This function will now be presented and then used in the sequel. It is defined by

$$g(\dot{\lambda}) = \frac{a_o}{a_1 + \exp\{-a_2 F(\dot{\lambda})\}} \;, \quad (3.35)$$

where

$$F(\dot{\lambda}) \triangleq F(\dot{\eta}) = \sinh\{a_3\dot{\eta} + a_3/2\} \;; \quad (3.36)$$

here $a_o$, $a_1$, $a_2$, and $a_3$ denote constants, and the *normalized interfilamentary velocity* $\dot{\eta}$ is defined by

$$\dot{\eta} = \frac{\dot{\lambda}}{-\bar{\dot{\lambda}}_o} = \xi \, \frac{\bar{\dot{\lambda}}}{-\bar{\dot{\lambda}}_o} \;, \quad (3.37)$$

where the CE is considered to be at optimum length $\bar{\lambda}$ and the parameter $\bar{\dot{\lambda}}_o$ denotes the (negative) maximum shortening velocity of the myostructures at length $\lambda = \bar{\lambda}$. Note that in the present model shortening velocities have negative values.

We now return to Equation (3.15) and substitute into it $g(\dot{\lambda})$, as defined by (3.35). We obtain for the externally observable contractile force $f^C$

$$f^C = q(\xi,\gamma) k(\xi) \frac{a_o}{a_1 + \exp\{-a_2 F(\dot{\eta})\}} - a_5[2 - \exp\{a_6'(\xi-1)\}]. \quad (3.38)$$

We shall consider all of the following observations on the muscle fibre to have been made at optimum length $\lambda = \bar{\lambda}$, i.e. at $\xi = 1$, so that $k(\xi) = 1$.

Since the resting fibre, when released from a shorter length, always extends itself to its rest length (where $\xi = 1$) (Ramsey and Street, 1940) and the resting value of the active state is $q_o$, it follows from (3.12) and (3.14), and from the condition $f^C(q_o,\bar{\xi}) = 0$, that

$$a_5 = q_o \bar{f}^{CP} = q_o \bar{f}/(1-q_o),$$

where $\bar{f}^{CP}$ denotes the isometric value of $f^{CP}$ at $\xi = 1$ and $\bar{f}$ is the isometric force output of the combined CE and sarcolemma PE under maximal stimulation and at $\lambda = \bar{\lambda}$, i.e. at $\xi = 1$. Hence (3.38) becomes

$$f^C = q(\xi,\gamma)k(\xi) \frac{a_o}{a_1 + \exp\{-a_2 F(\dot{\eta})\}} - \frac{q_o}{1-q_o} \bar{f}[2 - \exp\{a_6'(\xi-1)\}]. \quad (3.39)$$

For most mammalian muscles the constant $a_6'$ has a value of about 5.10 - 6.97 (estimated from passive fibre tensions of Fig. 6 of Edman, 1979).

If the non-stimulated fibre ($q = q_o$) is stationary ($\dot{\eta} = 0$), its force output $f^C$ at $\xi = 1$ is zero. On the other hand, the fully stimulated fibre ($q \approx 1$) shortening at maximum velocity ($\dot{\eta} = -1$) at normalized length $\xi = 1$ also produces zero force output (by definition of the maximum shortening velocity). These two conditions enable us to determine the constants $a_o$ and $a_1$ in (3.39) so that the relative force output $f^C/\bar{f}$ is now given by

$$f^C/\bar{f} = \frac{q(\xi,\gamma)k(\xi)}{b_2[a_1+\exp\{-a_2 F(\dot{\eta})\}]} - b_1[2 - \exp\{a_6'(\xi-1)\}], \quad (3.40)$$

where we have put

$$b_1 = q_o/(1-q_o), \quad b_2 = (1-q_o)/(a_1+\exp\{-a_2\sinh(a_3/2)\}), \quad (3.41)$$

and where $a_1$ is given by

$$a_1 = \frac{q_o\exp\{a_2\sinh(a_3/2)\} - \exp\{-a_2\sinh(a_3/2)\}}{1 - q_o} \quad (3.42)$$

If the fully stimulated fibre is extended at increasing velocities, its force output approaches a maximum asymptotically (see, for example, Sugi, 1972). Let this relative maximum stretching force be denoted by $\hat{f}^C/\bar{f}$. Then for $\dot{\eta}$ positive (stretching) and very large, we have that $F(\dot{\eta})$ is large, so that for $a_2 > 0, \exp\{-a_2 F(\dot{\eta})\}$ tends to zero. Under these conditions we find from (3.40) that

$$\hat{f}^C/\bar{f} = \frac{1}{a_1 b_2} - b_1, \quad (3.43)$$

which enables us to express the constant $a_2$ as

$$a_2 = \frac{-\ell n\{q_o(1-\bar{f}/\hat{f}^C)\}}{2\sinh(a_3/2)}. \quad (3.44)$$

Relation (3.40) corresponds to the force-velocity relation of classical muscle physiology. The relation is, however, obviously not identical with the velocity-dependence function $g(\lambda)$ as expressed by Equation (3.35). For a given experimentally observed force-velocity curve, with all the measurements performed near optimum length $\bar{\lambda}$, and at maximum stimulation ($q \approx 1$), the parameters $a_1$, $a_2$, $a_3$, $b_1$, and $b_2$ fitting the function (3.40) to the experimental data can be obtained as follows.

From the measurements on the fibre the following data are available: the values of $\bar{f}$, $\hat{f}^c$, and a number of points on the observed force-velocity curve. In addition, a value of $q_o = 0.003\text{-}0.005$ (preferably 0.005) is appropriate for most mammalian muscle fibres. A straightforward method of estimating the parameters is to guess a value of $a_3$ (between 2 and 5), and then compute $a_2$, $a_1$, $b_1$ and $b_2$ (in that sequence) from Equations (3.44), (3.42), and (3.41) respectively, until a satisfactory fit of the experimental data points is obtained. Alternatively, more sophisticated least-square methods for parameter identification can be used. These will be discussed in Chapter 5.

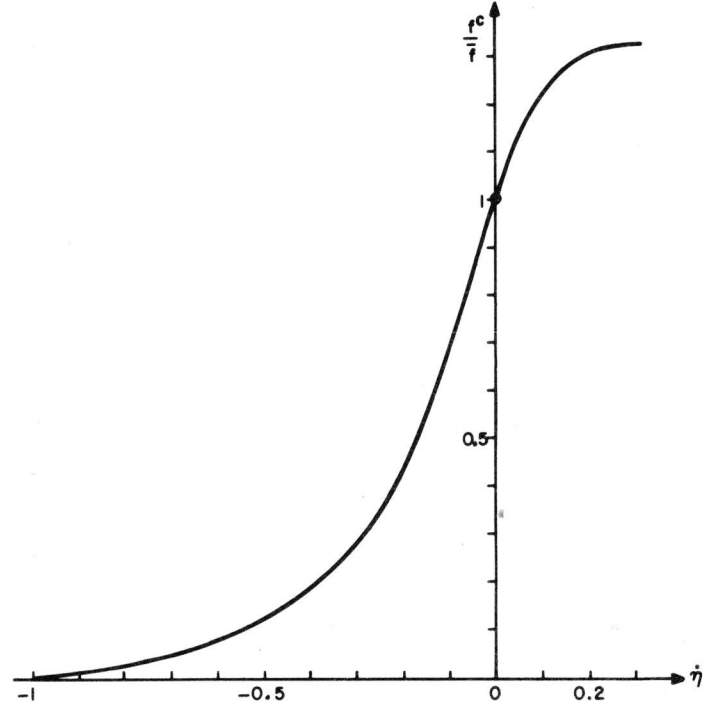

**FIG. 3.5** Normalized force-velocity curve for fast rat muscle fibre at 37°C. The parameter values providing the best fit for Equation (3.40) are $a_1 = 0.1073$, $a_2 = 1.409$, $a_3 = 3.2$, $b_1 = 0.005025$, $b_2 = 6.9828$, $q_o = 0.005$. The asymptote is at $\hat{f}^c/\bar{f} = 1.33$

It is interesting to note that in mammalian fast and slow muscles the normalized maximum shortening velocity may change by as much as a factor of three, while the curvature of the normalized force-velocity curve remains the same for both muscle types (Close, 1964).

A typical normalized force-velocity curve for a fast rat muscle fibre at 37°C is shown in Fig. 3.5. Note the change in curvature at $\dot{\eta} = -0.08$.

It should be noted that the present model correctly predicts the behaviour of the muscle fibre in slow to moderately fast stretches, i.e. for a velocity domain where practically all other known force-velocity models break down.

Equation (3.40) defines what may be termed the *monomorphic part of the universal volume* and is illustrated by Fig. 3.6.

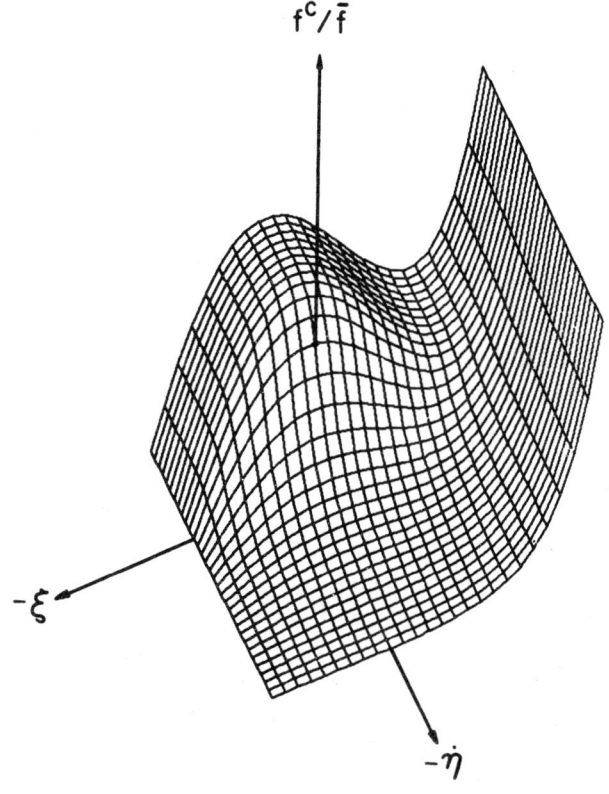

**FIG. 3.6:** The monomorphic part of the universal volume for the parameter value $q = 1$. The origin of the coordinate system is at $\xi = 1$, $\dot{\eta} = 0$ and $f^c/\bar{f} = 0$

This part of the universal volume is called monomorphic since no morphogenesis in the sense of catastrophe theory (Thom, 1975) takes place in this region, i.e. no discontinuity in the nature of the system occurs which would give rise to a change in its previous form. This region coincides with Hill's (1950b) region of 'thermodynamically reversible' changes in the contractile proteins, which region antecedes that of 'thermodynamically irreversible' changes. The universal volume encompasses all possible contractive states of the muscle fibre and has the phase plane $(\xi,\dot{\eta})$ as its base (we recall that $\dot{\eta} = \dot{\xi}\bar{\lambda}/(-\bar{\lambda}_o)$). The volume is generated by a sequence of evolving characteristic surfaces, each of which is defined by $\gamma$ = constant in Equation (3.40). Hence, as $\gamma(t)$ varies with time, so does the volume, which can be imagined to pulsate in the direction of the vertical axis according to the rhythm prescribed by the time change of the active state q.

Equation (3.40) may now be solved for $\dot{\eta}$, and by virtue of relations (3.36) and (3.37) we obtain the differential equation

$$\dot{\xi} = \frac{-\bar{\lambda}_o}{\bar{\lambda}}\left[\frac{1}{a_3}\text{arcsinh}\{-\frac{1}{a_2}\ell n\ \{\frac{q(\xi,\gamma)k(\xi)}{b_2[f^C/\bar{f} + b_1(2-\exp\{a_6'(\xi-1)\})]} -a_1\}\}-\frac{1}{2}\right],$$

(3.45)

with initial value $\xi_o$ and $f^C=f^{SE}$, where $f^{SE}/\bar{f}$ is given by (4.42) in Chapter 5.

This equation, which describes the *contraction dynamics* of the fibre, taken together with the *excitation system* (3.27), constitutes the complete set of state equations which describe the dynamic behaviour of the CE of a single fibre. The state variables are the free Ca ion concentration $\gamma$, and the normalized length $\xi$ of the contractile element CE. The system is controlled by the relative stimulation rate $v$.

If not the average response of the fibre but its detailed response to a train of nerve volleys is required, then (3.21-3.23) together with (3.29) and (3.45) must be used.

It is interesting to observe that (3.45) correctly predicts the dependence of the maximum shortening velocity $\dot{\eta}_o$ on the active state q and the length $\xi$ of the CE. By using (3.37) in (3.45) and putting $f^C = 0$ we find

$$\dot{\eta}_o(q,\xi) = \frac{1}{a_3}\text{arcsinh}\{-\frac{1}{a_2}\ell n\ \{\frac{qk(\xi)}{b_1b_2[2-\exp\{a_6'(\xi-1)\}]} -a_1\}\} - \frac{1}{2}. \quad (3.46)$$

Putting $\xi = 1$ and varying q only, the predicted active state dependence of $\dot{\eta}_o$ is obtained.

This relationship is illustrated by the left-hand diagram of Fig. 3.7. As can be seen, the maximum shortening velocity $\dot{\eta}_o$ is only slightly dependent on the active state q for values of q greater than 0.30. This is in complete agreement with the experimental finding of Thames et al. (1974) and Edman (1979) that $\dot{\eta}_o$ declines only by a few percent when q is varied between 0.30 and unity.

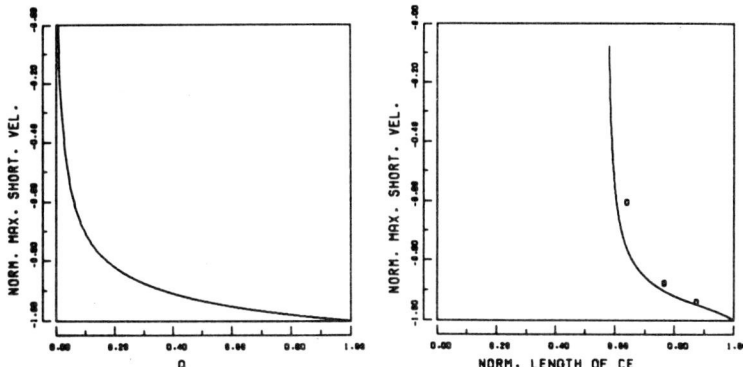

**FIG. 3.7:** Dependence of the normalized maximum shortening velocity $\dot{\eta}_o$ on the active state q (graph on the left) and the length $\xi$ of the contractile element (graph on the right). Solid lines: model predictions; symbols: experimental results. The values used for the constants in Equation (3.46) were those given in the legend of Fig. 3.5. For detailed explanations see the text

The right-hand graph of Fig. 3.7 displays the solution of (3.46) when q = 1 and $\xi$ is varied. Again, the prediction is compared with experimental results of Gordon et al. (1966, Fig.11) and Edman (1979), and a reasonable agreement is apparent.

The correct prediction by the present model of the length and active state dependence of the maximum shortening velocity constitutes a feature of the present model which is not equalled by any other so far evolved.

Before concluding this chapter we shall briefly discuss a peculiar phenomenon which occurs when a tetanized muscle fibre is stretched. In this contractive mode the force output of the fibre rises rather abruptly at first, and then continues to rise at a slower rate. Upon termination of the stretch the force decays rapidly initially, and then at a much slower rate. This gives rise to the well-known phenomenon of force enhancement after stretch (Déléze, 1961; Sugi, 1972; Hill, 1977; Flitney and Hirst, 1978; Edman et al., 1978; etc.), which increases with increasing fibre length beyond the optimum length $\ell$. This effect of stretch potentiation is most pronounced in frog muscle fibres at 0°C, but

49

is much less apparent in mammalian muscle at 37°C. A very detailed discussion of this phenomenon, including simulation results, is presented in Section 6.2.2 of Chapter 6.

This completes the derivation of the relations expressing the excitation and contraction dynamics of the contractile element of a single fibre. The next chapter will be devoted to the discussion of the transition from the distributed model displayed in Fig. 2.3 to the lumped model representing whole-muscle properties.

## CHAPTER 4

# A global myocybernetic control model of skeletal muscle

In the previous two chapters the mathematical formulation was established of the structural and functional properties of skeletal muscle. The basic elements have been shown to be incorporated in a distributed system which constitutes the muscle fibre. Under the assumption that the total muscle, comprising a specified number of fibres, interacts with an external mass, the equations of motion of this system will now be derived. However, although the distributed system is of value for some applications, it is generally too complex to be employed in the multilink system of the animal body. Thus a transition from the distributed system to a simpler lumped system will have to be made without sacrificing too many of the positive features of the more complex model. There is also another reason for such a transition: as will be seen, great difficulties are encountered when an attempt is made to define control parameters for the distributed model. And this, after all, is one of the main purposes of the present modelling procedure.

### 4.1 THE MODEL OF THE DISTRIBUTED SYSTEM

From Fig. 2.3 it follows that the total kinetic energy of the system is given by

$$T = (½)(m \sum_{i=1}^{\bar{\rho}} \sum_{j=1}^{2n} \dot{x}_{ij}^2 + M\dot{\ell}^2), \qquad (4.1)$$

where $\bar{\rho}$ is the number of fibres in the muscle, M is the external mass, and $\ell$, m, n and $x_{ij}$ are as indicated in Fig. 2.3 (n is obviously the number of sarcomeres in a typical fibre). By applying the Lagrangian formalism, the following equations of motion for the distributed system are obtained from (4.1) and the arrangement shown in Fig. 2.3:

$$m\ddot{x}_{i1} = f_{i1}^{CE} - f_{io}^{SE} + f_{i1}^{PE}, \qquad i=1,\ldots,\bar{\rho};$$

$$m\ddot{x}_{i,2k} = -f_{ik}^{CE} + f_{ik}^{SE} - f_{ik}^{PE}, \qquad i=1,\ldots,\bar{\rho};\ k=1,\ldots,n-1;$$

$$m\ddot{x}_{i,2k+1} = f_{i,k+1}^{CE} - f_{ik}^{SE} + f_{i,k+1}^{PE}, \qquad i=1,\ldots,\bar{\rho};\ k=1,\ldots,n-1;$$

$$m\ddot{x}_{i,2n} = -f_{in}^{CE} + f_{in}^{SE} - f_{in}^{PE}, \qquad i=1,\ldots,\bar{\rho};$$

$$M\ddot{\ell} = -F^{PE} - F^{SE}. \qquad (4.2)$$

In addition to (4.2) the following kinematic relations (see Fig. 2.3) also hold for $i = 1,\ldots,\bar{\rho};\ k = 1,\ldots,n$,

$$f_{ik}^{CE} = f_{ik}^{BE},$$

$$F_o^{SE} = \sum_{i=1}^{\bar{\rho}} f_{io}^{SE},$$

$$F^{SE} = \sum_{i=1}^{\bar{\rho}} f_{in}^{SE}, \qquad (4.3)$$

where $F_o^{SE}$ and $F^{SE}$ denote the total force across the series elastic structures at the origin and insertion of the muscle respectively.

As regards the forces $f_{ik}^{BE}$ across the series elastic elements of the cross-bridges, expression (3.8) derived in Section 3.2. is valid for a single cross-bridge only, while we are dealing here with whole fibre properties. The model (3.8) is, however, easily extended to account for all cross-bridges in a half-sarcomere of a fibre. Obviously, only cross-bridges attached to the actin-troponin-tropomyosin complex will contribute to this type of series elasticity. Their relative number $\Omega_{ik}$ for any contractive mode is given by (3.11), i.e. by

$$\Omega_{ik} = q(\xi_{ik},\gamma_i)k(\xi_{ik})H(\dot{\lambda}_{ik}),$$

where $H(\dot{\lambda}_{ik})$ expresses the velocity-dependence of $\Omega_{ik}$. At present, experimental evidence is insufficient to permit the identification with $H(\dot{\lambda}_{ik})$ of any particular function. However, in Chapter 5 we shall demonstrate that the most likely candidate is the function $h(\dot{\lambda}_{ik})=h(\dot{\xi}_{ik})$ as defined by (4.15), so that with $\bar{f}_{ik} = N\bar{f}$ and $\bar{f}$ defined by (3.8), the expression for $f_{ik}^{BE}$ becomes

$$f_{ik}^{BE} = \bar{f}_{ik} \, q(\xi_{ik}, \gamma_i) k(\xi_{ik}) h(\dot{\xi}_{ik}) \frac{\exp(\sigma \delta_{ik}) - 1}{\exp(\sigma) - 1}, \qquad (4.4)$$

where the damping term has been neglected (which is justified below), and where

$$\delta_{ik} \triangleq \lambda_{Bik}/\lambda_{Blik}, \qquad (4.5)$$

since $\lambda_{Boik} \equiv 0$.

It is noteworthy that the predicted dependence of $f_{ik}^{BE}$ on the degree $k(\xi_{ik})$ of filamentary overlap has recently been confirmed experimentally by Flitney and Hirst (1978, Fig. 8).

It should be emphasized that the present treatment of the distributed system dynamics is limited to the case of fusiform muscles, i.e. muscles in which the fibres are arranged in parallel. The model will, however, later be extended to include muscles with penniform fibre arrangement.

Since the elements $PE_{ik}$ represent the long-range stiffness of the sarcolemma, and the cross-bridge series elastic elements $BE_{ik}$ exhibit only short-range stiffness, the small length changes of the latter produce only insignificant force changes in the former elements. Hence the force output $f_{oik}(\lambda_{ik}, \lambda_{Bik})$ of $PE_{ik}$ can be closely approximated by the function $f_{oik}(\lambda_{ik})$ as defined by Equation (3.14). By a similar argument, the cross-bridge elasticity $BE_{ik}$ may be appended to the united element $CE_{ik}$ as shown in Fig. 4.1 below.

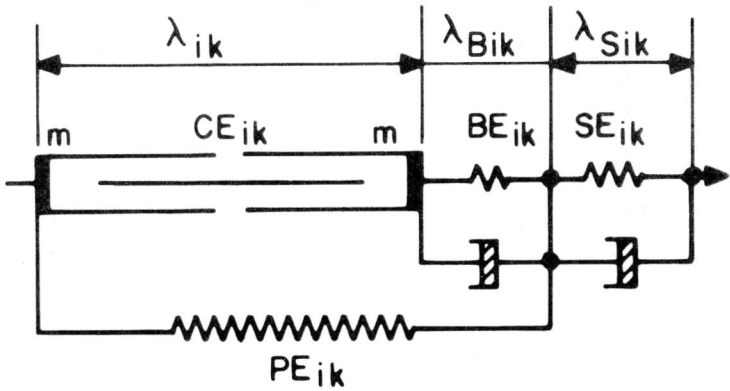

**FIG. 4.1:** Approximation of sarcomere model shown in Fig. 2.3. For explanation of symbols see the text

In accordance with the discussion in Chapter 3 (Equation (3.15)) the force $f_{ik}^{PE}$ produced by $PE_{ik}$ is already incorporated in the expression for the force output $f_{ik}^{C}$ of the contractile structure, and hence need no longer be considered separately. Also, in Section 3.2 it was pointed out that the damping of all the series elastic elements in the muscle is comparatively small. While this damping component plays a decisive role if a single muscle fibre contracts under zero load, it becomes negligible when the whole muscle in the intact animal body is considered. Thus for all practical purposes one may put $k_s' = 0$ in Equations (3.6), (3.7) and (3.8) if the total muscle is to be modelled.

With these assumptions, Equations (4.2) simplify somewhat. However, the fact remains that a large muscle may contain 400 and more motor units, each of which consists of 1 to 2 000 muscle fibres. If we assume an average length of 3 cm for the fibres we find that there are n = 12 000 sarcomeres in a fibre. Since the dynamics of each sarcomere is described by two differential equations, the system (4.2) would consist of 9 600 001 nonlinear second-order differential equations for this muscle, even though only whole motor units instead of single fibres were considered. This clearly illustrates that the distributed system of the form presented by (4.2) has theoretical value only and that a realistic approach necessitates a transition to a lumped model.

## 4.2 THE MODEL OF THE LUMPED SYSTEM

In addition to the assumptions made in the previous section we may justifiably assume that all fibres belonging to a motor unit are more or less identical and activated at approximately the same time.

Now, in the animal body a muscle never contracts under zero load. In general, the mass (or equivalent mass) it has to move is large when compared with its own mass. Hence under these conditions the internal mass of the muscle may be neglected and the force relationships for the distributed system result from simple kinematical equilibrium conditions by putting m ≡ 0 in (4.2). They are found to be (refer to Figs. 2.3 and 4.1)

$$f_{io}^{SE} = f_{ik}^{C} = f_{ik}^{SE} = f_{i,k+1}^{C} = f_{i,k+1}^{SE}, \quad f_{ik}^{CE} = f_{ik}^{BE}, \qquad (4.6)$$

for all k=1,...,n-1, and all i=1,...,p̄.

By virtue of Equation (3.7) with the damping factor $k_s'$ put equal to zero and with the relative extension δ defined by

$$\delta = (x_{i,2k+1} - x_{i,2k} - \lambda_{soik})/(\lambda_{slik} - \lambda_{soik}) \qquad (4.7)$$

for the k-th series elastic element $SE_{ik}$ shown in Fig. 2.3 the assumption of identical elements (Eqn. (4.6)), i.e.

$$f^{SE}_{ik} = f^{SE}_{i,k+1}, \; k=1,\ldots,n-2,$$

permits us to write

$$\exp\left\{\frac{\sigma}{\lambda_{slik}-\lambda_{soik}}(x_{i,2k+1}-x_{i,2k}-\lambda_{soik})\right\} =$$

$$= \exp\left\{\frac{\sigma}{n(\lambda_{slik}-\lambda_{soik})}[n(x_{i,2k+1}-x_{i,2k})-n\lambda_{soik}]\right\}. \tag{4.8}$$

Define

$$\lambda^*_{si} \triangleq n(x_{i,2k+1}-x_{i,2k}),$$

$$\lambda^*_{soi} \triangleq n\lambda_{soik}, \tag{4.9}$$

$$\lambda^*_{sli} \triangleq n\lambda_{slik}.$$

Then it follows from (4.8) that all series elastic elements $SE_{ik}$, k=1,...,n−1, of the i-th fibre can be replaced by an equivalent single element $SE_i$ of length $\lambda^*_{si}$, rest length $\lambda^*_{soi}$, and isometric maximum extension $\lambda^*_{sli}$. The same procedure can be applied to the elements $CE_{ik}$, $BE_{ik}$, $SE_{io}$ and $SE_{in}$, resulting in equivalent lumped elements $CE_i$, $BE_i$, and $\overline{SE}_i$ respectively. Renewed application of the procedure to the new series elastic elements $SE^*_i$ and $\overline{SE}_i$ results in a totally lumped series elastic element $SE_i$ for the i-th fibre, and the partially lumped model of the total muscle takes the form shown in Fig. 4.2.

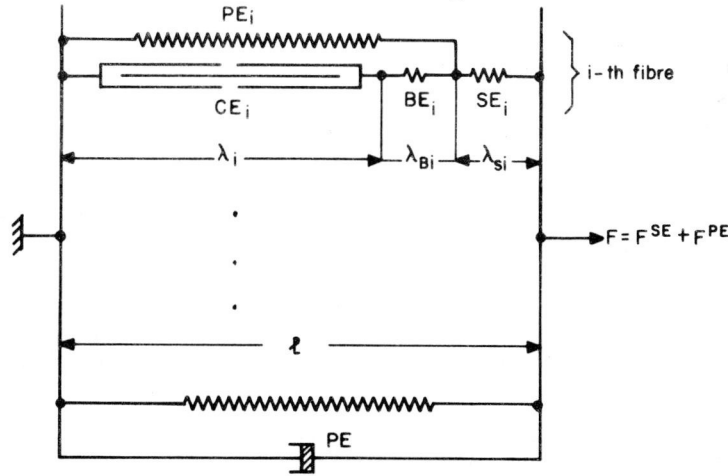

**FIG. 4.2:** Model of the semi-lumped system structure of the total muscle. For details refer to the text

It should be understood that the functional properties of the elements constituting the distributed system are completely transferable to the elements making up the lumped system. For instance, all equations describing the excitation and contraction dynamics of the CE of the single muscle fibre remain valid if we substitute $\lambda_i$ for $\lambda_{ik}$, $\bar{\lambda}_i$ for $\bar{\lambda}_{ik}$, $\dot{\lambda}_i$ for $\dot{\lambda}_{ik}$, and $\dot{\bar{\lambda}}_{oi}$ for $\dot{\bar{\lambda}}_{oik}$ in all expressions where these variables appear. This follows immediately from the fact that the length and contraction velocity of the lumped contractile element $CE_i$ are defined by

$$\lambda_i \triangleq n\lambda_{ik}$$

and

$$\dot{\lambda}_i \triangleq n\dot{\lambda}_{ik} \qquad (4.10)$$

respectively. Likewise, the expressions derived for the functional properties of the elements $BE_{ik}$ and $SE_{ik}$ of the distributed system remain valid also for the elements $BE_i$ and $SE_i$ of the partially lumped system if the appropriate transformations of the type (4.9) and (4.10) are carried out.

It must, however, be emphasized again that the lumping procedure leading to the macroscopic model rests on the assumption that the effect of the internal sarcomere mass elements becomes negligible when contractive modes are considered in which the external mass moved by the muscle is large compared with these mass elements. If this assumption cannot be made, the lumping procedure as carried out here becomes invalid, and the model of the distributed system must be used.

## 4.3 THE MYODYNAMICS OF FUSIFORM MUSCLES

We shall now derive the global myocybernetic control model for fusiform muscles, while the case of penniform muscles will be treated in the next section.

In fusiform muscles, all fibres are arranged approximately in parallel, as indicated in Fig. 4.2. The total force output F of the muscle is composed of the sum of all the series elastic forces $f_i^{SE}$ of the $\bar{\rho}$ fibres, and the force $F^{PE}$ across the parallel elastic element. Note that the forces across series elastic elements of non-stimulated fibres are, in general, not strictly zero since the resting fibre still has a non-zero active state $q_o$ (D.K. Hill, 1968). We therefore have the following force relations (see Fig. 4.2):

$$F = \sum_{i=1}^{\bar{\rho}} f_i^{SE} + F^{PE} = F^{SE} + F^{PE}, \qquad (4.11)$$

and

$$f_i^{SE} = f_i^C, \quad f_i^{CE} = f_i^{BE}, \quad i=1,\ldots,\bar{\rho}. \qquad (4.12)$$

From (4.12) it follows that

$$F^{SE} = \sum_{i=1}^{\bar{\rho}} f_i^{SE} = \sum_{i=1}^{\bar{\rho}} f_i^C = F^C. \qquad (4.13)$$

Using expression (3.40) for the force output $f_i^C$ of the contractile structure of the i-th fibre, relation (4.13) can be written as

$$F^C = \sum_{i=1}^{\bar{\rho}} \bar{f}_i \, [q(\xi_i,\gamma_i)k(\xi_i)h_i(\dot{\xi}_i) - b_{1i}(2 - \exp\{a'_6(\xi_i-1)\})], \qquad (4.14)$$

where

$$h_i(\dot{\xi}_i) \triangleq \frac{1}{b_{2i}[a_{1i}+\exp\{-a_{2i}F_i(\dot{\xi}_i)\}]}, \qquad (4.15)$$

with $F_i(\dot{\xi}_i) = F_i(\dot{\lambda}_i)$ defined by (3.36), all other symbols having the same meaning as in Chapter 3.

Expression (4.14) states that the total force output $F^C$ of a 'pseudo' contractile structure C is equal to the sum of the force outputs $f_i^C$ of the individual muscle fibres, each of which is (generally) in a different active state $q(\xi_i,\gamma_i)$, exhibits a different degree of filamentary overlap represented by $k(\xi_i)$, and shows a different velocity dependence $h_i(\dot{\xi}_i)$. With the large number of fibres present in a muscle there would be no hope of obtaining a usable expression for $F^C$ from (4.14).

Fortunately, things are not quite so complicated. First, and most important, the nervous system does not control single muscle fibres, but sets of fibres, which are grouped into so-called motor units. Each motor unit consists of the motoneuron producing the neural input signal to the fibres of that unit, the axon which conducts the signal, and the set of muscle fibres which are supplied by the branches of that axon (the feedback pathways are not discussed here). In general, the fibres of a motor unit are distributed randomly over a certain volume of the muscle (the motor unit territory), and all have similar morphological, contractile, and histochemical profiles (specific ATPase activity, degree of oxidative and glycolytic enzyme activity, mitochondrial content, etc.). A great contribution to muscle physiology was that by E. Henneman when he elucidated the realtionships between properties of the motoneuron, the axon (size, conduction velocity, etc.), and the fibres (size, number, histochemical profiles) belonging to a specific motor unit. His findings were summarized in the 'size-principle' (Mendell and Henneman, 1971). As we shall see presently, these findings have a profound influence on the modelling procedure to be adopted here.

From the above discussion it is clear that the summation in (4.14) should be changed so as to run over the number of motor units and not over the number $\bar{\rho}$ of fibres present in a muscle. We shall, however, go one step further. Instead of considering single motor units we shall group motor units into three types of dynamically different populations: the N-population of *stimulated* units; the R-population of units which were stimulated but had been switched off some time ago and are still active (so-called *semi-active* units); and the population of *inactive* (resting) units. Let the fibres of the i-th motor unit occupy a fraction $\triangle_i u$ of the total cross-sectional area occupied by all the fibres of that muscle. Since the maximum isometric tension per unit of fibre cross-sectional area (the intrinsic strength) is about the same for all fibre types (Close, 1972) we have that

$$\bar{f}_i = \bar{F} \triangle_i u, \tag{4.16}$$

where $\bar{F}$ denotes the maximum isometric force output of the whole muscle. For the three motor unit populations defined above, Expression (4.14) then becomes

$$F^c/\bar{F} = \sum_{i=1}^{N(t)} \triangle_i u [k(\xi_i) q(\xi_i, \gamma_i^+) h_i(\ddot{\xi}_i) - b_1 k_1(\xi_i)] +$$

$$+ \sum_{j=N(t)+1}^{N(t)+R(t)} \triangle_j u [k(\xi_j) q(\xi_j, \gamma_j^-) h_j(\ddot{\xi}_j) - b_1 k_1(\xi_j)] +$$

$$+ \sum_{k=N(t)+R(t)+1}^{\bar{N}} \triangle_k u [k(\xi^\circ) q_o h_k(\dot{\ell}) - b_1 k_1(\xi^\circ)], \tag{4.17}$$

where we have put $k_1(\xi) = 2 - \exp\{a_6'(\xi-1)\}$, and where $\bar{N}$ denotes the total number of motor units present in a particular muscle. The symbols $\gamma^+$ and $\gamma^-$ denote the Ca-concentrations of the stimulated and semi-active units respectively.

It should be clearly understood that the three populations of motor units are not defined by their anatomical locations within the muscle but rather by their state of excitation.

The severe difficulties bedevilling the modelling of the present situation can be appreciated when the following general contractive mode is considered: N units, all of different size and with different contractive properties, were activated at different times $t_i$ by different stimulation rates $v_i$. Hence the expressions

$$\Delta_i u[k(\xi_i)q(\xi_i,\gamma_i^+(v_i(t-t_i)))h_i(\dot{\xi}_i)-b_1k_1(\xi_i)]$$

in (4.17) are all different, nonlinear functions of time, delayed in their onset by $(t-t_i)$. If now $N(t)$ increases, the N-population absorbs part of the semi-active R-population, all units of which have different active states since they have been switched on and off at various previous times.

At first sight the prospects of obtaining a solution to this problem appear as bleak as for the case of single fibres. However, we shall demonstrate that by deriving certain average properties of the three motor unit populations, it is possible to reduce the lumped model to one contractile element CE in series with the corresponding series elastic elements BE and SE, and including the element PS (which represents the renamed parallel elasticity of the sarcolemmas), and one parallel elastic element PE.

Before this can be done, certain characteristics of mammalian motor units and their associated recruitment dynamics must be specified.

By now it has been fairly well established (Desmedt and Godaux, 1977; Grimby and Hannerz, 1977; Henneman and Olson, 1965; Henneman, 1968; Henneman et al., 1965; Milner-Brown et al., 1973b, etc.) that motor units are recruited in a sequential order, according to their sizes, in all types of non-pathological contractions (static isometric, isometric ramp, reflex, brisk ballistic) and that the cumulative relative cross-sectional area u occupied by the fibres of the recruited units increases (Hatze, 1979; Hatze and Buys, 1977) approximately according to

$$u = u_o \exp(N\bar{c}/\bar{N}), \quad 0 < u_o \le u \le 1, \qquad (4.18)$$

where the constant $\bar{c}$ is related to $u_o$ by $\bar{c} = -\ell n u_o$; N denotes the number of stimulated motor units; and $\bar{N}$ is the total number of units present in a specific muscle. Note that for $\bar{N}$ large, (4.18) is sufficiently accurate for the present purpose. Exactness is achieved by replacing N by (N–1).

The increment $\Delta_i u$ of the relative cross-sectional area u upon recruitment of the i-th unit follows from (4.18) as

$$\Delta_i u = A \exp(\bar{c}i/\bar{N}), \qquad (4.19)$$

where the constant A is determined by the requirement $\sum_{i=1}^{\bar{N}} \Delta_i u = 1$, i.e.

$$A = 1 / \sum_{i=1}^{\bar{N}} \exp(\bar{c}i/\bar{N}). \qquad (4.20)$$

As is easily verified by differentiation and subsequent discretization, the law (4.18) implies that

$$\Delta_i u / u_{i-1} = (\bar{c}/\bar{N}) \Delta N, \quad i=1,\ldots,\bar{N}, \qquad (4.21)$$

which, for unit increment of $\Delta N = 1$ (i.e. an increment of one motor unit), represents Weber's law for motor unit recruitment or the *size law for motor units* (Hatze, 1979). It is interesting to note that this law can be regarded as a consequence of a much more general principle of minimum transentropy in biological information processing systems (Hatze, 1979).

Since $u_{i-1}$ is the cumulative of the increments $\Delta_i u$, i.e.

$$u_{i-1} = u_o + \sum_{j=1}^{i-1} \Delta_j u,$$

we obtain, by using (4.16),

$$u_{i-1} = \bar{f}_o/\bar{F} + \sum_{j=1}^{i-1} \bar{f}_j/\bar{F} = \sum_{j=0}^{i-1} \bar{f}_j/\bar{F}. \qquad (4.22)$$

By substituting (4.16) and (4.22) into (4.21), and putting $\Delta N = 1$, we find that

$$\bar{f}_i / \sum_{j=0}^{i-1} \bar{f}_j = \bar{c}/\bar{N} = \text{constant}. \qquad (4.23)$$

Relation (4.23) states that the ratio of the tetanic (i.e. maximum isometric) tension of the i-th motor unit, to the cumulative tetanic tension of all previously recruited units should be (approximately) constant. That this is indeed the case, except for very small motor units where sampling bias could have distorted the results, has been demonstrated experimentally by Henneman and Olson (1965, Table 4). Similar findings were also made by Milner-Brown *et al.* (1973b).

We have thus specified two very important properties of motor-unit recruitment dynamics: motor units are normally recruited sequentially from the smallest to the largest (and deactivated in reverse order), and the 'size' of the recruited units grows exponentially according to (4.19).

The paper by Henneman and Olson (1965, Tables 2 and 3) contains other valuable information which enables us to deduce additional properties of motor units. If the contraction times $t_{ci}$ of motor units are plotted as functions of the corresponding tetanic tensions $\bar{f}_i$, the best least-squares fit of all the functions tried is obtained with the model

$$t_c = a_o - a_1' \ell n \bar{f}, \qquad (4.24)$$

where $a_o$ and $a_1'$ are constants. This result held true for the motor units of both the cat soleus and medial gastrocnemius muscles. Using (4.16) in (4.21), and dropping the index i we find

$$t_c = a_o - a_1' \ell n(\bar{c}\bar{F}u/\bar{N}) = a_o' - a_1' \ell n u = a_2' - a_3' N/\bar{N}, \qquad (4.25)$$

where u has been substituted from (4.18), and $a_2'$, $a_3'$ are constants.

Expression (4.25) means that the *contraction time $t_c$ of a motor unit is a decreasing linear function of the order number N of that unit.*

For further reference let us introduce the *normalized number n of recruited motor units* defined by

$$n \triangleq N/\bar{N}, \qquad (4.26)$$

where we shall regard n as a continuous variable. This is a reasonable approximation for $\bar{N}$ large.

With (4.26), Equation (4.25) can be written as

$$t_c = a_2' - a_3' n. \qquad (4.27)$$

Good estimates of the constants $a_2'$ and $a_3'$ are usually available. For n = 0, the value of $t_c$ corresponds to the contraction time $t_{co}$ of the slowest unit in that muscle, while for n = 1, the value of $t_{c1}$ corresponds to the fastest contracting unit. Hence $a_2' = t_{co}$ and $a_3' = t_{co} - t_{c1}$. For human muscles, extreme values are about $t_{co} = 0.140$ s and $t_{c1} = 0.030$ s (small hand muscles), so that $a_2' = 0.14$ and $a_3' = 0.11$.

Relation (4.27) has another interesting implication. Close (1965, p. 544) showed that for mammalian skeletal muscle the normalized speed of shortening $(-\bar{\lambda}_o/\bar{\lambda})$ (see Equation (3.45)) is related to the contraction time $t_c$ of a muscle consisting predominantly of one fibre type, by

$$(-\bar{\lambda}_o/\bar{\lambda}) = B/t_c, \qquad (4.28)$$

where the constant B has values of 0.297, 0.280, and 0.294 respectively for slow human, fast cat, and slow cat muscle. Using (4.27) we can write

$$(-\bar{\lambda}_o/\bar{\lambda}) = B/(a_2'-a_3'n) . \qquad (4.29)$$

This means that a fairly good estimate of the value of $(-\vec{\lambda}_o/\bar{\lambda})$ can be obtained once the values of $a_2'$ and $a_3'$ are known for a particular muscle. For values of B = 0.297, $a_2' = 0.14$, and $a_3' = 0.11$, we find that $(-\vec{\lambda}_o/\bar{\lambda}) = 2.12$ s$^{-1}$ and 9.90 s$^{-1}$ (approximately) for the slowest (n = 0) and fastest (n = 1) human motor units respectively.

Another important parameter also depends on the contraction time $t_c$ of a motor unit. It is the parameter m appearing in Equation (3.27). This parameter is a rate constant and as such is inversely related to $t_c$, i.e.

$$m = A_o/t_c = A_o/(a_2'-a_3'n) = 1/(A_2-A_3n) , \qquad (4.30)$$

where $A_o \approx 0.372$ for human muscle, and the constants $A_2$ and $A_3$ are defined by

$$A_2 \triangleq a_2'/A_o, \quad A_3 \triangleq a_3'/A_o. \qquad (4.31)$$

Hence for muscles containing motor units composed of different fibre types, the parameter m becomes the function m(n) given by (4.30).

Finally, the maximum firing rate $\hat{v}$ of a motor unit is also related to the contraction time $t_c$. This relation (details in Hatze, 1978a) is given by

$$\hat{v} = k'm = k'A_o/t_c = A_o'/(a_2'-a_3'n), \qquad (4.32)$$

where k' denotes a constant.

The predicted function $\hat{v}(t_c) = A_o'/t_c$ has, in fact, recently been observed experimentally by Grimby, Hannerz and Hedman (1979, Fig. 3) on motor units of the human short toe extensor muscle. The authors also conclude that the voluntary discharge frequency range of a motor unit can be used as an indication of its contraction time.

We have established a number of important relationships describing properties of mammalian motor units. These relationships will now be used in the construction of a global myocybernetic control model of skeletal muscle.

As a first step we postulate the existence of a normalized 'average' length $\xi$ of the completely lumped contractile element CE (see Fig. 4.3). The above postulate will be substantiated in Appendix A1. It implies that the model now takes the form shown in Fig. 4.3.

From Fig. 4.3 we see that the following relations hold:

$$F^{CE}/\bar{F} = F^{BE}/\bar{F} , \; F^{SE}/\bar{F} = F^C/\bar{F} = (F^{PS}+F^{BE})/\bar{F} , \qquad (4.33)$$

where $F^{PS}$ denotes the renamed parallel elastic force which is due to sarcolemma elasticity.

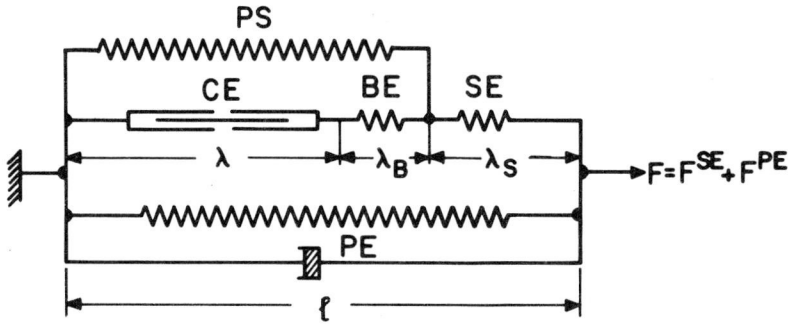

**FIG. 4.3:** Model of the completely lumped system structure of the total muscle. Explanations in the text

The forces $F^{CE}$ and $F^{BE}$ designate 'average' forces, in some sense, the exact definitions of which have to be established. To this end, let us rewrite (4.17) with $\xi_i \equiv \xi_j \equiv \xi^o \equiv \xi$. Then

$$F^C/\bar{F} = k(\xi) \left[ \sum_{i=1}^{N(t)} \triangle_i uq(\xi,\gamma_i^+)h_i(\dot{\xi}) + \sum_{j=N(t)+1}^{N(t)+R(t)} \triangle_j uq(\xi,\gamma_j^-)h_j(\dot{\xi}) + \sum_{k=N(t)+R(t)+1}^{\tilde{N}} \triangle_k uh_k(\dot{\xi})q_o \right] - b_1 k_1(\xi) \quad , \qquad (4.34)$$

since $\sum_{i=1}^{\tilde{N}} \triangle_i u = 1$.

Comparing the structure of (4.34) with that of one of the three sums in (4.17) we see that certain similarities exist. Indeed, if functions $\varepsilon(\cdot)$ and $h(\dot{\xi},.)$ could be found such that

$$\varepsilon(\cdot)h(\dot{\xi},.) = \sum_{i=1}^{N(t)} \triangle_i uq(\xi,\gamma_i^+)h_i(\dot{\xi}) + \sum_{j=N(t)+1}^{N(t)+R(t)} \triangle_j uq(\xi,\gamma_j^-)h_j(\dot{\xi}) + q_o \sum_{k=N(t)+R(t)+1}^{\tilde{N}} \triangle_k uh_k(\dot{\xi}) \quad , \qquad (4.35)$$

the model structure would remain unchanged.

In general, we cannot hope to find functions $\varepsilon$ and $h$ which satisfy (4.35) *exactly*. However, it is possible to derive expressions for $\varepsilon$ and $h$

which closely approximate (4.35). Indeed, our objective will be to derive model equations which

(i) preserve the model structure;
(ii) contain no more sums; and
(iii) contain directly or indirectly a control parameter which reflects motor unit recruitment.

More specifically, we want to find 'pseudo Ca-concentrations' $\psi(\cdot)$ and $\varphi(\cdot)$ such that the *excitation function* $\varepsilon$ can be expressed as

$$\varepsilon = A[\, q(\xi,\psi) \sum_{i=1}^{N(t)} \exp(\bar{c}i/\bar{N}) + q(\xi,\varphi) \sum_{j=N(t)+1}^{N(t)+R(t)} \exp(\bar{c}j/\bar{N}) +$$

$$+ q_o \sum_{k=N(t)+R(t)+1}^{\bar{N}} \exp(\bar{c}k/\bar{N})] \qquad (4.36)$$

where (4.19) and (4.20) have been used.

The sums appearing on the right-hand side of Expression (4.36) can all be replaced by analytical terms which do not contain any summations. This replacement has not been carried out here, in order to make the transition from (4.17) to (4.36) more transparent. It can also be seen that the model structure is completely preserved. Moreover, the new control parameter representing motor unit recruitment will appear in the differential equations describing the dynamics of the variables $\psi$ and $\varphi$.

As the derivation of the differential system describing the recruitment, excitation and contraction dynamics of the global model is rather involved, it has been transferred to Appendix A1. We shall give here only the final result. The state equations for the global myocybernetic control model are found to be

$$\dot{n} = \hat{n}z, \quad 0 \le n \le 1, \qquad n(0) = n_o,$$

$$\dot{r} = -\hat{n}z \frac{r - w^-\bar{\delta}}{r + \bar{\delta}} - (1+w^-) \frac{m(n,r)r}{10^{-3}m(n,r) + (\varphi/k_2c)^2}, \qquad r(0) = r_o,$$

$$\dot{\psi} = m(n)\,[cv-\psi] + w^+z\bar{c}\hat{n}\,\frac{1 - \exp\{\rho_o(\xi)(\psi-\varphi)\}}{\rho_o(\xi)(1-\exp\{-\bar{c}n-\bar{\delta}\})} - (1+w^-)m(n,r)\varphi,$$
$$\psi(0) = \psi_o,$$

$$\dot{\varphi} = -m(n,r)\varphi - w^-\,[m(n)\varphi\,(\frac{cv}{\psi+\bar{\delta}}-1) - z\bar{c}\hat{n}\,\frac{1-\exp\{\rho_o(\xi)(\varphi-\psi)\}}{\rho_o(\xi)(\exp\{\bar{c}r+\bar{\delta}\}-1)}]\,, \qquad \varphi(0) = \varphi_o,$$

$$\dot{\xi} = \frac{1}{S}\left[\frac{1}{a_3}\operatorname{arcsinh}\left\{-\frac{1}{a_2}\ell n\left\{\frac{k(\xi)\varepsilon}{b_2[F^{SE}/\bar{F}+b_1k_1(\xi)]}-a_1\right\}\right\}-\tfrac{1}{2}\right], \quad \xi(0) = \xi_o,$$
(4.37)

where the functions $m(n,r)$, $m(n)$, $\rho_o(\xi)$, $S$, and $\varepsilon$ are given repectively by Equations (A1.56), (A1.39), (A1.15), (A1.77), and (A1.73), with the associated expressions (A1.70)–(A1.72), (A1.27), (A1.46), (A1.69) in Appendix A1, and where $k_1(\xi)=2-\exp\{a_6'(\xi-1)\}$. The constants $n_o$, $r_o$, $\psi_o$, $\varphi_o$, $\xi_o$ denote the initial values of the respective state variables, while $\delta$ is a small constant with a value of about $10^{-8}$. The muscle-specific constant $\hat{n}$ designates the maximum rate of motor unit recruitment, $k_2c$ has a fixed value of $1.373 \times 10^{-8}$ (for explanation see Appendix A1). The constants $\bar{c}$, $c$, $a_1$, $a_2$, $a_3$, $b_1$, and $b_2$ have the meaning defined previously.

The symbols $w^+ = w^+(z)$ and $w^- = w^-(z)$ denote switching functions that switch parts of the state equations on and off according to the value of the control $z$. They are defined by

$$\begin{aligned}w^+(z) &= 1, & w^-(z) &= 0, & \text{for } z &> 0, \\ &= 0, & &= 0, & \text{for } z &= 0, \\ &= 0, & &= -1, & \text{for } z &< 0.\end{aligned}$$
(4.38)

The state variables $n$, $r$, $\psi$, $\varphi$, and $\xi$ denote respectively the normalized populations of stimulated (n) and semi-active (r) motor units, the 'pseudo' Ca-concentrations of stimulated ($\psi$) and semi-active ($\varphi$) populations, and the normalized length of the lumped contractile element.

The *control parameter* $z$, $-\dot{z} \leq z \leq 1$ ($\dot{z}$ a positive constant), appears here for the first time and designates the *normalized rate of motor unit recruitment*. If $z = 1$, then motor units are recruited at the maximum rate $\hat{n}$ ($\hat{n} \approx 14 - 20$ s$^{-1}$ for large human muscles). If $z = 0$, no recruitment takes place, while $z < 0$ indicates derecruitment.

The *control parameter* $v$, $0 \leq v \leq 1$, denotes the *normalized average stimulation rate* of the whole n-population (see Equation (3.28) for the single fibre) and is defined by

$$v(t) = \frac{\int_0^{\bar{c}n} v(x,t)\exp(x)dx}{\exp(\bar{c}n) - 1} \qquad (4.39)$$

where, by (3.28) and (4.32),

$$v(x,t) = 1/(\hat{v}(x)\tau(x,t)) = (a_2' - a_3'x/\bar{c})/A_o'\tau(x,t), \qquad (4.40)$$

where $\tau(x,t) = \tau(\bar{c}i/\bar{N},t)$ is the discharge interval at time $t$ of the i-th motor unit in the n-population of stimulated units. Obviously, Relation

(4.39) defines a weighted average stimulation rate where the weights are the motor unit sizes which grow exponentially with n.

If we now compare Equations (3.27) and (3.45) for the single fibre with the system (4.37) describing the global dynamics of the whole muscle, we can clearly see the changes introduced by the inclusion in the model of a varying number of stimulated motor units having varying biochemical and contractile profiles: the *recruitment dynamics* of the global model is given by the first two differential equations of system (4.37), while the *excitation dynamics,* previously defined for the single fibre by the γ-dynamics (3.27), has its counterpart in the dynamics of the pseudo Ca-concentrations ψ and φ of the stimulated and semi-active motor units respectively. Finally, the contraction dynamics (3.45) for the single fibre is completely analogous to the *contraction dynamics* of the stimulated and semi-active units of the global model, as represented by the last of the equations of the system (4.37).

By virtue of Fig. 4.3, the total relative force output of the global model is now given by

$$F/\bar{F} = F^{PE}/\bar{F} + F^{SE}/\bar{F} , \qquad (4.41)$$

where

$$F^{SE}/\bar{F} = [\exp\{\frac{\sigma}{\bar{\alpha}\lambda_{so}}(\ell - \bar{\lambda}(\xi+\bar{\kappa}) - \lambda_{so})\} - 1] / [\exp(\sigma) - 1] , \qquad (4.42)$$

and where the expression (4.45) below for $\lambda_B$ has been used. As will be demonstrated in Chapter 5, the constant $\bar{\kappa}$ may be set equal to zero, without loss of generality.

It should be noted that the function $F^{SE}/\bar{F}$, which appears in the last of Equations (4.37), is precisely the function defined by Equation (4.42) above.

For a given length $\ell$ of the *resting* muscle, the initial value $\xi_o$ of the normalized length of the contractile element can be found by solving the nonlinear equation

$$F^{SE}(\ell,\xi_o)/\bar{F} = b_1[k(\xi_o) - k_1(\xi_o)] , \qquad (4.43)$$

for $\xi_o$, where $k_1(\xi_o)$ has been defined after (4.37). If $\ell < \bar{\lambda}(1+\bar{\kappa})+\lambda_{so}$, the inactive muscle is below its rest length. It will, however, be in static equilibrium, which implies $\xi_o = 1$. In this case (4.42) yields $F_o^{SE}/\bar{F} < 0$, which is unrealistic since a tendon cannot support compressive forces. Hence we put $F_o^{SE}/\bar{F} = 0$, which then also satisfies (4.43).

The first term on the right-hand side of Equation (A1.78) in Appendix A1 represents the normalized contractile force $F^{CE}/\bar{F}$. By virtue of the first of Equations (4.33) this is equal to $F^{BE}/\bar{F}$, the structure of which expression is represented by (4.4). As can be seen, this equality implies that

$$[\exp\{\sigma\lambda_B/\bar{\kappa}\bar{\lambda}\}-1]/[\exp\{\sigma\}-1] -1 = 0. \tag{4.44}$$

Equation (4.44) has the solution $\lambda_B = \bar{\kappa}\bar{\lambda}$, which gives the important result

$$\lambda_B = \bar{\kappa}\bar{\lambda} = \text{constant}, \tag{4.45}$$

with $\bar{\kappa} \approx 0.03 - 0.06$ for mammalian skeletal muscle. This result will be discussed more fully in Chapter 5.

Relations (4.45) implies that $\bar{\kappa}$ becomes an additive constant in (4.42). By putting $\bar{\kappa} \equiv 0$, a *redefinition* of $\xi$ can be achieved which satisfies the arrangement shown in Fig. 2.3.

## 4.4 THE MYODYNAMICS OF PENNIFORM MUSCLES

Let it be assumed that the fibres of a penniform muscle, whose common tendon runs vertically, are inclined to the horizontal at varying angles $\theta_i$, $i=1,\ldots,\bar{\rho}$. In addition, the fibres are assumed to have individual tendinous parts, lie on a general, curved surface, and have different lengths, as is the case with living muscle in situ. Since the contractive forces are transmitted along the centroids of the fibres, the biomechanically equivalent representation shown in the left-hand part of Fig. 4.4 is valid. Note that only the left-hand part of the fibre population is shown, as the fibres are assumed to be symmetrically distributed about the vertical.

It is our objective to find the model of a lumped element whose reponses closely approximate the responses of the distributed system consisting of a multitude of different fibres. Such a lumped model is shown in the right-hand part of Fig. 4.4.

The required lumped model must have the property that

$$F^{SE}\sin \theta(\ell,\lambda_T) \approx \sum_{i=1}^{\bar{\rho}} f_i^{SE} \sin \theta_i, \tag{4.46}$$

for the whole range of possible values of $\ell$ and $\lambda_T$, and for all contractive states of the muscle. We shall use the following notation (see also Fig. 4.4). The actual muscle length, from the origin of the longest fibre to the insertion of the common tendon, is denoted by L. The length $\bar{L}$ is that length at which the penniform muscle develops the maximum isometric force $\bar{F}^{ST}$ across the common tendon. The tendon has length $\lambda_T$ at non-zero force, and rest length $\lambda_{T_0}$. The i-th muscle fibre is inclined to the horizontal at an angle $\theta_i$, has a contractile element of length $\lambda_i$, an

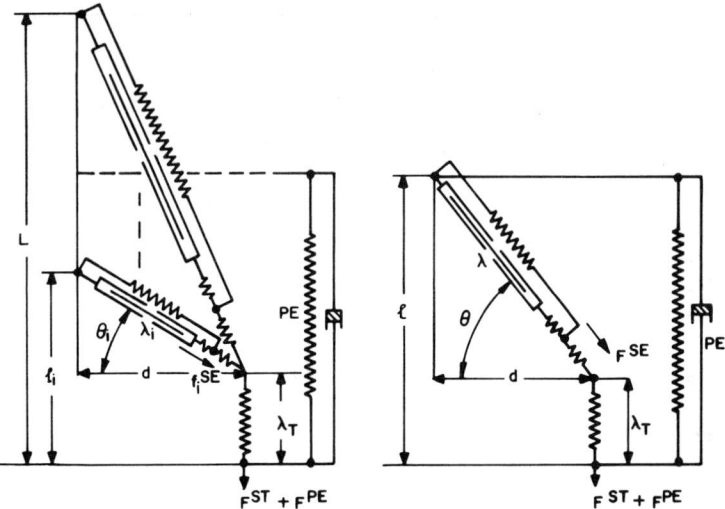

**FIG. 4.4:** Biomechanically equivalent representation of penniform muscles (left-hand diagram), and corresponding lumped model (rigth-hand diagram). Detailed explanations in the text

elastic element of length $\lambda_{Bi}$ which accounts for cross-bridge elasticity (not shown in Fig. 4.4), a series elastic element of length $\lambda_{si}$, and produces an instantaneous force output $f_i^{SE}$. The inclination to the horizontal of the longest fibre is $\hat{\theta}$, while that of the shortest fibre is $\check{\theta}$. The muscle also contains a parallel elastic element PE. The notation for the lumped model shown in the right-hand part of Fig. 4.4 is analogous to that of the left-hand part.

The equations describing the lumped model are derived in Appendix A2 and will only be summarized here. Let the lumped model muscle be at length $\bar{\ell}$ (which corresponds to $\bar{L}$ in the real muscle), and completely relaxed. Then $\theta = \theta_o$, where $\theta_o$ is given (see (A2.17)) by

$$\theta_o = \arctan\{(\ell n \cos \check{\theta}_o - \ell n \cos \hat{\theta}_o)/(\hat{\theta}_o - \check{\theta}_o)\}, \qquad (4.47)$$

i.e. the resting inclination $\theta_o$ of the lumped model is computed from the resting inclinations $\check{\theta}_o$ and $\hat{\theta}_o$ of the shortest and longest fibres of the real muscle respectively.

By applying simple trigonometry we find from Fig. 4.4 that

$$\bar{\ell} = \lambda_{To} + d \tan \theta_o = \bar{L} - (\tan \hat{\theta}_o - \tan \theta_o)d, \qquad (4.48)$$

$$\bar{\theta} = \arctan\{\tan \theta_o - \bar{\alpha}\lambda_{To}/d\}, \qquad (4.49)$$

$$\sin \theta(\ell,\lambda_T) = (\ell - \lambda_T)/(d^2 + (\ell - \lambda_T)^2)^{1/2}. \qquad (4.50)$$

Also
$$\bar{\lambda} = \frac{d/\cos\bar{\theta} - (1+\bar{\alpha})\lambda_{so}}{1 + \bar{\kappa}}, \quad (4.51)$$

where (see (A2.11))
$$\lambda_{so} = \bar{\lambda}/k_s = \bar{\lambda}/(\bar{k}_s - \sigma\bar{\kappa}\sigma_s^2/2\bar{\alpha}), \quad (4.52)$$

i.e. the 'average' rest length $\lambda_{so}$ of the series elastic element of the lumped model can be computed from the mean ratio $\bar{k}_s$ and the variance $\sigma_s^2$ of the probability distribution of the individual fibres (for details see Appendix A2).

Since estimates for $\bar{L}$, $\lambda_{To}$, and $\hat{\theta}_o$ of the real muscle are available, the value of d can be found from

$$d = (\bar{L} - \lambda_{To})/\tan\hat{\theta}_o . \quad (4.53)$$

Finally, the total force output of the model muscle is given by
$$F = F^{ST} + F^{PE},$$

where
$$F^{ST} = \bar{F}\sin\bar{\theta}[\exp\{\sigma(\lambda_T/\lambda_{To}-1)/\bar{\alpha}\}-1]/[\exp(\sigma)-1], \quad (4.54)$$

which denotes the force across the common tendon, and where $\bar{F}$ is the maximum isometric force in the direction of $F^{SE}$ (see Fig. 4.4).

On the other hand we have that
$$F^{ST} = F^{SE}\sin\theta(\ell,\lambda_T), \quad (4.55)$$

where $\sin\theta(\ell,\lambda_T)$ is given by (4.50), and $F^{SE}/\bar{F}$ is defined by Equation (4.42), $\ell$ being replaced by $[d^2+(\ell-\lambda_T)^2]^{1/2}$, where $\ell$ in this expression refers to the length of the lumped model as shown in Fig. 4.4.

In addition, as was the case with fusiform muscles, we may put $\bar{\kappa} \equiv 0$ in (4.42) which, in effect, means that the constant structure $\bar{\kappa}\bar{\lambda}$ is incorporated into the CE of total length $\bar{\lambda}\xi$.

From (4.54) and (4.55) we see that a new variable, $\lambda_T$, has entered the equations describing the force output of the muscle. This means that at each instant t, the nonlinear equation

$$F^{ST}(\lambda_T(t)) = F^{SE}(\ell(t),\xi(t),\lambda_T(t))\sin\theta(\ell(t),\lambda_T(t))$$

(see (4.54) and (4.55)) must be solved for the unknown $\lambda_T(t)$.

Sometimes it is found that the tendinous parts of the individual fibres are so short that they may be ignored when compared with the elasticities residing in the common tendon. In this case, the expression $F^{SE}/\bar{F}$ in (4.37) must be replaced by $F^{ST}/(\bar{F}\sin\theta)$,

where
$$\sin \theta = [1 - (d/(\bar{\lambda}\xi))^2]^{1/2},$$
$$\lambda_T = \ell - [(\bar{\lambda}\xi)^2 - d^2]^{1/2}, \qquad (4.56)$$

and $F^{ST}$ is given by (4.54) with $\lambda_T$ defined by (4.56). The total force output of the model muscle then follows from (4.41) with $F^{SE}/\bar{F}$ replaced by $\bar{F}^{ST}/\bar{F}$.

If this case applies, no nonlinear equation needs to be solved, since $\lambda_T$ is now given directly by (4.56).

The excitation and contraction dynamics for penniform muscles is, of course, identical with that of fusiform muscles, and is given by (4.37).

However, in the expression for $F^{SE}/\bar{F}$, which appears in the last of Equations (4.37), and which is defined by (4.42), the variable $\ell$ has to be replaced by $[d^2 + (\ell - \lambda_T)^2]^{1/2}$.

This completes the discussion of the myodynamics of penniform muscles.

CHAPTER 5

# Methods for model parameter estimation

A global control model of skeletal muscle was developed in the previous chapter. This model was termed 'myocybernetic control model' in order to emphasize its myocybernetic nature, i.e. the fact that its control parameters correpond to the real neural controls stimulation rate and rate of motor unit recruitment.

If a model is to be useful in practical applications it is essential that its parameter values should be identifiable. Thus methods are required which will make it possible for the user of the model, from experimental observations made on the real biosystem, to estimate the values of the parameters (other than the control parameters) that characterize the model. This is the so-called parameter identification problem.

For skeletal muscle *in situ* the externally observable output quantities are: the torque created by the contractive and passive myostructures acting against some bony leverage, and the surface accelerometermyogram (Lammer *et al.*, 1976). In some sense, the morphometric characteristics of the muscle (muscle length and mass, coordinates of origin and point of insertion of the muscle, etc.) could also be classified as observable output quantities.

The surface (or indwelling-wire) electromyogram represents an observable which is indicative of the neural *control input* to the muscle, and hence is not an output quantity in the above sense.

The set of parameters that characterize the global model, and the values of which are to be determined, may be divided into the subsets of *myodynamic* and *myocybernetic* parameters. The set $\{P_{md}\}$ of myodynamic parameters is given by

$$\{P_{md}\} = \{\bar{F}, s_k, \bar{\alpha}, \bar{\ell}, \bar{L}, \lambda_{so}, \lambda_{To}, \sigma, \bar{\kappa}, \bar{c}, c, q_o, a_3, \hat{F}^C/\bar{F}, a_2', a_3', \check{\xi}, s, k_2, \bar{\delta}, A_o, B, \check{\theta}_o,$$
$$\hat{\theta}_o, k_s\}, \tag{5.1}$$

while the set $\{P_{mc}\}$ of myocybernetic parameters is defined by

$$\{P_{mc}\} = \{\hat{n}, \check{z}, A_o'\} . \tag{5.2}$$

For convenience we shall restate the meaning of these parameters.

The symbol $\bar{F}$ denotes the maximum isometric force of the muscle, measured at full stimulation, at optimum length $\bar{\ell}$, and in the direction of force production of the CE.

The parameters $s_k, \bar{\alpha}, \bar{\ell}, \bar{L}, \lambda_{so}$ and $\lambda_{To}$ denote respectively the spread of the Gaussian-like filamentary-overlap function $k(\xi)$ as defined by (3.34); the relative maximum isometric extension of the series-elastic element and defined by

$$\bar{\alpha} = (\lambda_{s1} - \lambda_{so})/\lambda_{so}; \tag{5.3}$$

the optimum muscle length at which the maximum isometric force $\bar{F}$ is produced; the real optimum length of penniform muscles (see Fig. 4.4); the resting length of the total series-elastic element; and the resting length of the tendinous part of the series-elastic element.

The parameter $\bar{\alpha}$ is obviously related to the parameter

$$\bar{\xi} = (\lambda_{s1} - \lambda_{so})/\bar{\ell}, \tag{5.4}$$

which designates the normalized maximum isometric extension of the tendinous SE relative to the optimum muscle length $\bar{\ell}$, and which was introduced in Section 3.2. Since $\bar{\ell} = (1 + \bar{\kappa})\bar{\lambda} + \lambda_{s1}$, we find from (5.3) and (5.4) that

$$\bar{\alpha} = \frac{\bar{\xi}\,\bar{\ell}}{(1-\bar{\xi})\bar{\ell} - (1+\bar{\kappa})\bar{\lambda}} , \tag{5.5}$$

where we may put $\bar{\kappa} \equiv 0$ in accordance with the discussion in Chapter 4. For fusiform muscles the value of $\lambda_{so}$ can be found from

$$\lambda_{so} = 0.98\,\lambda_{To} + 0.02\,\bar{\ell} , \tag{5.6}$$

if the tendon rest length $\lambda_{To}$ is known. Equation (5.6) follows from the fact that the Z-discs occupy about 2% of the length of the contractile structures.

For penniform muscles, the rest length $\lambda_{so}$ of the tendinous parts of the individual muscle fibres is frequently negligible when compared with the rest length $\lambda_{To}$ of the common tendon of all fibres. If the assumption $\lambda_{so} \approx 0$ cannot be made, then $\lambda_{so}$ is determined by (4.52), or must be estimated from the geometry of the muscle.

Further parameters appearing in (5.1) are the constant $\sigma$ which determines the degree of nonlinearity of the stress-strain relation of the series-elastic elements, the recruitment-range constant $\bar{c}$ as defined after (4.18), the maximum sarcoplasmic calcium concentration c (see (3.27)), the resting active state $q_o$, the force-velocity constant $a_3$, the relative maximum contractive force $\hat{F}^C/\bar{F}$ as defined by (3.43), the constants $a_2'$ and $a_3'$ which relate to the contraction times of the motor units (see (4.27)), the projected upper limit $\hat{\xi}$ of the range of possible normalized contractile element lengths, and the exponent s as defined by (3.30) and (3.32).

The constants $k_2$, $\bar{\delta}$, $A_o$ and B appearing in (5.1) all have fixed values which will be given below. The symbols $\hat{\theta}_o$ and $\check{\theta}_o$ denote respectively the resting inclinations of the shortest and longest fibre of a penniform muscle, while the ratio $k_s$ is defined in Appendix A2.

The set (5.2) of myocybernetic parameters consists of the maximum recruitment rate $\hat{n}$, the maximum rate of derecruitment $\check{z}$, and the constant $A_o'$ as defined by (4.32).

Some parameters that appear in the model equations but are not included in (5.1) or (5.2), can be computed from the latter. These are: $a_1$, $a_2$, $b_1$ and $b_2$, which are given by (3.42), (3.44) and (3.41) for those values of $a_3$ and the resting active state $q_o$ which provide the best fit for the normalized force-velocity curve as depicted in Fig. 3.5; the optimum length of the CE for fusiform muscles (where we may put $\bar{\kappa} \equiv 0$)

$$\bar{\lambda} = [\bar{\ell} - (1+\bar{\alpha})\lambda_{so}]/[1+\bar{\kappa}]$$
$$= [ (0.98 - 0.02\bar{\alpha})\bar{\ell} - 0.98(1+\bar{\alpha})\lambda_{To}]/[1+\bar{\kappa}] ; \qquad (5.7)$$

the corresponding value $\bar{\lambda}$ as given by (4.51) for penniform muscles; and the constants $\theta_o$, $\bar{\theta}$, and d which follow from (4.47), (4.49) and (4.53) respectively. The latter three constants are computed by using estimates of $\hat{\theta}_o$, $\check{\theta}_o$, and $\bar{L}$ (for details see (4.47)). In addition, the constants $A_2$ and $A_3$ appearing in (4.30) are defined by (4.31) in terms of $A_o$, $a_2'$, and $a_3'$.

For a given value of $\hat{F}^C/\bar{F}$, the values of the constants $a_1$, $a_2$, $a_3$, $b_1$, $b_2$, and $q_o$ characterizing the normalized force-velocity relation can be determined by the iterative algorithm outlined in Section 3.3.2. For the convenience of the reader we present in Table 5.1 the set of values of $a_1$, $a_2$, $a_3$, $b_1$, $b_2$ and $q_o$ providing the best fit to the normalized force-velocity curve for certain given values of $\hat{F}^C/\bar{F}$.

As mentioned previously, some of the parameters appearing in (5.1) and (5.2) are constants with fixed values. These values have been collected in Table 5.2.

The value of the constant $A_o'$ appearing in (4.32) has been estimated from data on ballistic contractions of the human short toe extensor muscles (Grimby *et al.*, 1979).

**TABLE 5.1:** Values of constants characterizing the normalized force-velocity curve

| $\hat{F}^C/\bar{F}$ | $a_1$ | $a_2$ | $a_3$ | $b_1$ | $b_2$ | $q_o$ |
|---|---|---|---|---|---|---|
| 1.15 | 0.13227291 | 0.71768166 | 4.8 | 0.0030090271 | 6.5568671 | 0.003 |
| 1.25 | 0.12712674 | 0.95230510 | 4.0 | 0.0050251256 | 6.2677356 | 0.005 |
| 1.33 | 0.10727021 | 1.4085387 | 3.2 | 0.0050251256 | 6.9828297 | 0.005 |
| 1.45 | 0.08797751 | 1.6983625 | 2.8 | 0.0050251256 | 7.8119211 | 0.005 |
| 1.60 | 0.07253144 | 1.9598826 | 2.5 | 0.0050251256 | 8.5899743 | 0.005 |

**TABLE 5.2:** Values of fixed myodynamic and myocybernetic constants for human skeletal muscle

| $\sigma$ | $\bar{\kappa}$ | $c$ | $s$ | $k_2$ | $\bar{\delta}$ | $A_o$ | $B$ | $A_o'$ |
|---|---|---|---|---|---|---|---|---|
| 1.531 | 0.0306 | $1.373 \times 10^{-4}$ | 1 | $10^{-4}$ | $\approx 10^{-8}$ | 0.372 | 0.297 | $\approx 3.60$ |

We have now identified a number of parameter values of the sets (5.1) and (5.2). Eliminating these from the respective sets, we are left with those parameters whose values must be estimated by means of experimental methods. The reduced set of myodynamic parameters is then given by

$$\{\bar{F}, s_k, \bar{\alpha}, \bar{\ell}, \bar{L}, \lambda_{To}, \bar{c}, \hat{F}^C/\bar{F}, a_2', a_3', \check{\xi}, \hat{\theta}_o, \hat{\theta}_o, k_s\}, \quad (5.8)$$

while the reduced set of myocybernetic parameters is

$$\{\hat{n}, \check{z}\}. \quad (5.9)$$

The experimental methods to be employed will consist of morphometric, myodynamic and electromyographic techniques. We shall begin by describing the morphometric procedures.

### 5.1 MORPHOMETRIC METHODS

Estimates of the values of the paramters $\bar{\ell}, \bar{L}, \lambda_{To}, \check{\theta}_o, \hat{\theta}_o$, and $k_s$ can be obtained by applying morphometric techniques, although in the case of $\bar{\ell}$ and $\bar{L}$ additional methods have to be employed. Where measurements on cadaver material must be made, the preparations should be fresh or well preserved.

In general, relative relations are established whenever possible. For instance, the rest length $\lambda_{To}$ of a muscle tendon is advantageously expressed as a proportion of the rest length of the total muscle. In some cases it is even possible to express the estimate of a morphometric parameter value in terms of a limb-segment dimension which can be observed externally on a living subject. This is clearly the most desirable procedure.

The values of $\bar{\ell}$ and $\bar{L}$ (for penniform muscles) cannot in general be determined by applying morphometric procedures only. By definition, $\bar{\ell}$, ($\bar{L}$) represents that length of the fusiform (penniform) muscle at which the maximum isometric force output is produced. Hence myodynamic measurements of maximum isometric torque outputs must be made at a sequence of muscle lengths (see Section 5.2.1). However, in order to evaluate these measurements, the actual length $L(\phi_j)$ of the muscle, and the muscle moment arm $D(\phi_j)$ must be known for each angular position $\phi_j$, $j=1,\ldots,p$, of the joint spanned by the muscle (or muscle group) under investigation. In general, both $L(\phi)$ and $D(\phi)$ are complicated and nonlinear functions of the joint angle $\phi$, owing to irregular bone structure and the non-stationarity of instantaneous centres of rotation of human joints. In most cases, $L(\phi)$ can be represented as

$$L(\phi) = L' + \triangle L(\phi), \qquad (5.10)$$

where $L'$ denotes the shortest *in situ* length of the muscle, and $\triangle L(\phi)$ is the length increment due to changes of $\phi$. The value of $L'$ can usually be expressed as a proportion of some segment dimension (for instance the arm length), while the functions $\triangle L(\phi)$ and $D(\phi)$ can be estimated from a sequence of X-ray photographs of the joint in question. How this is done will be demonstrated below by means of an example. With the functions $L(\phi)$ and $D(\phi)$ known, the value of $\bar{\ell}$ (or $\bar{L}$) can be estimated by the computational method discussed in Section 5.2.1.

The values of $\lambda_{To}$, $\hat{\theta}_o$, $\hat{\theta}_o$ and $k_s$ are estimated as average values from cadaver material. Examples are given below.

As an illustration of the above techniques we shall outline the procedure of estimating the values of those parameters of the human triceps muscle that can be determined by morphometric techniques.

The triceps muscle consists of three separate parts (the medial, lateral, and long heads) which are all penniform and insert into a common tendon. The tendon, in turn, has its insertion on the olecranon of the ulna, as shown in Fig. 5.1.

At each joint angle $\phi$, the distance $D(\phi)$ is the normal to the force vector **F** of the triceps muscle. The distance $r(\phi)$ is measured from the instantaneous centre of rotation of the joint to the *effective* point of force application of the tendon.

The humero-ulnar joint is a hinge joint with an approximately constant centre of rotation. This fact considerably simplifies the determi-

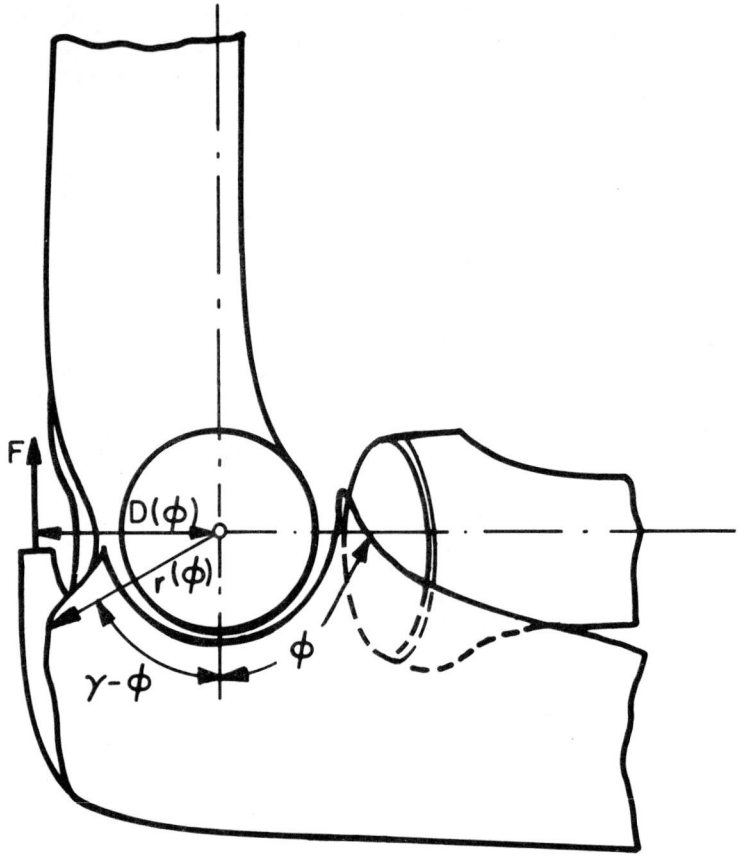

**FIG. 5.1:** Tracing of an X-ray photograph of the right elbow joint of a male subject aged 23. The symbols are explained in the text

nation of the functions $\triangle L(\phi)$ and $D(\phi)$ for the triceps muscle. The procedure is as follows. Nine successive X-ray photographs of the elbow joint flexed at 0° (full extension), 15°, 30°,...,120°, and with the triceps muscle moderately contracted are taken in the saggital plane (see Fig. 5.1). A metal marker of known length is placed on the skin beneath the ulna to provide a length scale in the X-ray photographs. Appropriately scaled readings of the values $D(\phi_j)$, $r(\phi_j)$, and $\gamma(\phi_j)$ (see Fig. 5.1), j=1,...,9, are then taken from the successive X-ray photographs, where $\phi_1=0,\ldots, \phi_9=2.0944$ rad (120°). The values of $D(\phi_j)$, $r(\phi_j)$ and $\gamma(\phi_j)$ are plotted on graph paper, and a decision is made on physical grounds as to which mathematical function is to be fitted to the experimental points. The parameters of this function are then determined by least-squares methods using the given data points.

In the present case, trigonometric considerations require that

$$D(\phi) = r(\phi)\sin(\gamma(\phi)-\phi), \quad \text{for } D(\phi) \geq r_1,$$
$$= r_1, \quad \text{for } D(\phi) < r_1, \tag{5.11}$$

where the model equations for $r(\phi)$ and $\gamma(\phi)$ are respectively given by

$$r(\phi) = r_1 + (1+c')(r_o-r_1)/[c' + (2+c')^{\frac{2\phi}{\pi}}], \tag{5.12}$$

and

$$\gamma(\phi) = 2.007, \quad \text{for } 0 \leq \phi \leq 1.047,$$
$$= 0.96 + \phi, \quad \text{for } 1.047 < \phi \leq 2.62. \tag{5.13}$$

A study of seven cadaver arms provided good average estimates for the values of $r_o = r(\phi=0)$ and $r_1 = r(\phi=2.62)$ in terms of the forearm length $\ell_{fa}$ (measured from the lateral epicondyle to the wrist joint centre). The estimates are

$$r_o \approx 0.129\,\ell_{fa}, \quad r_1 \approx 0.47\,r_o,$$

dimensions being in metres. The best-fitting value of the parameter $c'$ was found to be 5000.

The length increment $\triangle L(\phi)$ of the muscle is given by

$$\Delta L(\phi) = \int_0^\phi r(\psi)d\psi. \tag{5.14}$$

With the above expressions for $c'$, $r_o, r_1$, and $\gamma(\phi)$, relations (5.11) and (5.14) can be respectively written as

$$D(\phi) = \ell_{fa}\,[0.0606 + 341.92/(5000 + 5002^{\frac{2\phi}{\pi}})]\sin(\gamma(\phi)-\phi),$$
$$= 0.0606\,\ell_{fa}, \quad \text{for } D(\phi) < 0.0606\,\ell_{fa}, \tag{5.15}$$

and

$$\Delta L(\phi) = \ell_{fa}[0.129\phi + 0.01261\,\ell n\,\{5001/(5000 + 5002^{\frac{2\phi}{\pi}})\}], \tag{5.16}$$

where the dimensions are m and rad.

For a value of $\ell_{fa} = 0.239$ the functions $D(\phi)$ and $\triangle L(\phi)$ are shown in Fig. 5.2.

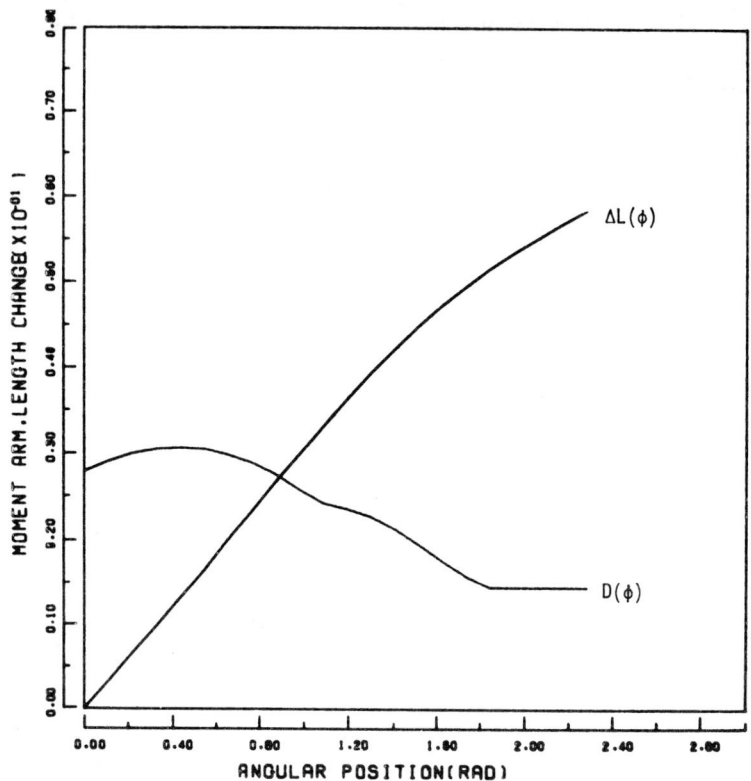

**FIG. 5.2:** Muscle moment arm function $D(\phi)$ and muscle length change $\triangle L(\phi)$ for the triceps muscle. The irregularity in the function $D(\phi)$ near $\phi = 1.05$ rad is due to the proximal protrusion on the olecranon (see Fig. 5.1)

Since the triceps muscle consists of the medial (muscle 1), lateral (muscle 2), and long head (muscle 3), the functions $L_i(\phi), i=1,2,3$, will differ by constant values determined by the different shortest *in situ* lengths $L_i'$. Similarly, the values of the tendon rest lengths $\lambda_{Toi}, i=1,2,3$, are all different because the tendons of the three heads are of different lengths, although morphologically they do combine to a common structure. On the other hand, the values of $\check{\theta}_{oi}$, $\hat{\theta}_{oi}$, and $k_{si}$, $i=1,2,3$, dot not appear to be significantly different for the three heads.

$L_1' \approx 0.43 \, \ell_a, \quad L_2' \approx 0.60 \, \ell_a, \quad L_3' \approx 0.32 \, \ell_a,$

$\lambda_{To1} \approx 0.12 \, \ell_a, \quad \lambda_{To2} \approx 0.14 \, \ell_a, \quad \lambda_{To3} \approx 0.12 \, \ell_a,$

$\check{\theta}_{oi} \approx 1.13, \qquad \hat{\theta}_{oi} \approx 1.40, \quad k_{si} = \bar{\lambda}_i/\lambda_{soi} \approx 12\text{–}165, \quad i=1,2,3, \qquad (5.17)$

where $\ell_a$ denotes the length of the upper arm, and where the comparatively large values of $k_{si}$ indicate that the rest lengths $\lambda_{soi}$ of the tendinous parts of the muscle fibres are negligible when compared with the rest lengths $\lambda_{Toi}$ of the tendons proper (see Fig. 5.3).

This completes the discussion of the morphometric procedures used to estimate certain parameter values. In the next section we shall discuss the myodynamic methods.

## 5.2 MYODYNAMIC METHODS

The set (5.8) of myodynamic parameters whose values are still to be estimated has been reduced further by the elimination of those parameters whose values can be determined by morphometric techniques. The reduced set is given by

$$\{\bar{F}, s_k, \bar{\alpha}, \bar{\ell}, \bar{L}, \bar{c}, \hat{F}^C/\bar{F}, a_2', a_3', \check{\xi}\}. \tag{5.18}$$

The present section will be devoted to the discussion of the methods used to determine these parameter values. Basically, the procedures can be categorized into those involving observations of maximum *steady-state* isometric torque outputs at *various* muscle lengths, and those involving observations of *dynamic* isometric torque outputs at a *fixed* muscle length. We shall begin with the former.

### 5.2.1 Observations of maximum steady-state isometric torque outputs at various muscle lengths

We shall treat the more general case of penniform muscles, among which the fusiform muscles represent a special case. Also, it will be assumed that a number m ($1 \leq m \leq 10$) of muscles, all of which have different myodynamic characteristics, contribute collectively to the contractile torque observed externally on a limb.

Since we are dealing with *maximum-effort* (implying maximal neural stimulation), *steady-state, isometric* contractions, we have that in (4.37)

$$\dot{n} = \dot{r} = \dot{\psi} = \dot{\varphi} = \dot{\xi} = 0, \tag{5.19}$$

and

$$n = 1, \; r = \varphi = 0, \; \psi = c, \; \varepsilon \approx 1,$$

for each of the m penniform muscles involved. Obviously, the dynamic problem reduces to a static one.

By virtue of (4.37), (3.41), and (3.34), the conditions $\dot{\xi}_i = 0$ and $\varepsilon_i \approx 1$, i=1,...,m, imply

$$F_i^{SE}/\bar{F}_i = \exp\{-(\frac{\xi_i-1}{S_{ki}})^2\}/(1-q_{oi})-b_{li}[2-\exp\{6.97(\xi_i-1)\}], \quad (5.20)$$

where $F_i^{SE}/\bar{F}_i$ is given by (4.42), with $\ell_i$ replaced by

$$[d_i^2 + (\ell_i - \lambda_{Ti})^2]^{1/2}, \quad (5.21)$$

so that

$$F_i^{SE}/\bar{F}_i = [\exp\{\frac{\sigma}{\bar{\alpha}_i\lambda_{soi}}([d_i^2+(\ell_i-\lambda_{Ti})^2]^{1/2}-\bar{\lambda}_i\xi_i-\lambda_{soi})\}-1]/[\exp\{\sigma\}-1], \quad (5.22)$$

where we have put $\bar{\kappa} \equiv 0$ in accordance with the discussion in Chapter 4.

In addition, (4.55) holds, with $F^{ST}$ given by Equation (4.54). When (5.22) is substituted into (4.55) it is seen that the resulting equation can be solved for $\xi_i$, yielding

$$\xi_i = [[d_i^2+(\ell_i-\lambda_{Ti})^2]^{1/2}-\lambda_{soi}[1+\frac{\bar{\alpha}_i}{\sigma}\ell n\{1+\frac{\sin\bar{\theta}_i}{\sin\theta(\ell_i,\lambda_{Ti})}[\exp\{\sigma(\lambda_{Ti}/\lambda_{Toi}-1)/\bar{\alpha}_i\}-1]\}]]/\bar{\lambda}_i, \quad (5.23)$$

where $\sin\theta(\ell_i,\lambda_{Ti})$ is given by (4.50).

As discussed in the previous section, it is frequently possible to put $\lambda_{soi} = 0$, in which case (5.23) simplifies considerably.

By virtue of (5.20), Equation (4.55) can be written as

$$\frac{\sin\bar{\theta}_i}{\sin\theta(\ell_i,\lambda_{Ti})}[\exp\{\sigma(\lambda_{Ti}/\lambda_{Toi}-1)/\bar{\alpha}_i)\}-1]/[\exp\{\sigma\}-1] =$$

$$= \exp\{-(\frac{\xi_i-1}{S_{ki}})^2\}/(1-q_{oi})-b_{li}[2-\exp\{6.97(\xi_i-1)\}], \quad (5.24)$$

where $\xi_i(\lambda_{Ti})$ is defined by (5.23), and the model muscle length $\ell_i$ (see Fig. 4.4) is related to the real length $L_i$ of the i-th penniform muscle by (4.48), i.e. by

$$\ell_i = L_i - (\tan\hat{\theta}_{oi} - \tan\theta_{oi})d_i. \quad (5.25)$$

Thus (5.24) constitutes a nonlinear equation in the unknown $\lambda_{Ti}$. It can be solved numerically for a given value of $\ell_i$ (or $L_i$) and given values of the parameters appearing in the equation.

The further discussion will be facilitated by referring to Fig. 5.3.

From Figs. 4.4 and 5.3 it is clear that if there are m penniform muscles acting across a joint, the total *predicted* externally observable maximum *output torque* T produced by this group is given by

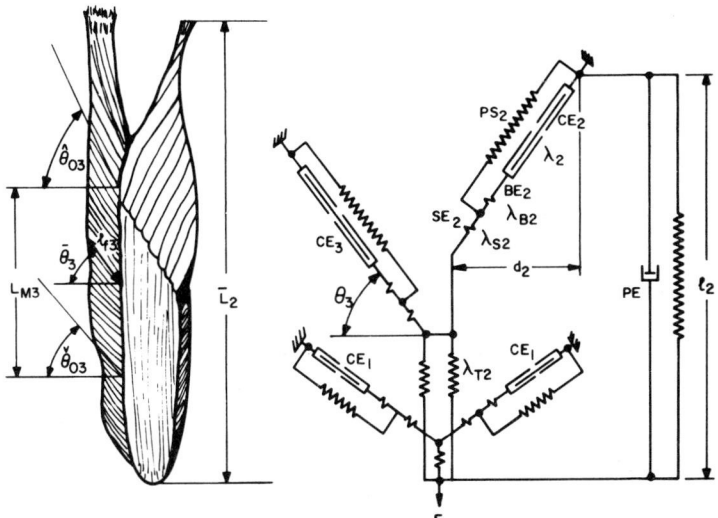

**FIG. 5.3:** Diagrammatic representation of a group of penniform muscles (human triceps muscle). Sketch of anatomical arrangement (left) and corresponding equivalent biomechanical representation (right). The symbols are explained in the text

$$\tilde{T} = \sum_{i=1}^{m} D_i(\phi)[F_i^{ST}(\phi) + F^{PE}(\phi)], \quad (5.26)$$

where $D_i(\phi)$ is the muscle moment arm of the i-th muscle at the angular position $\phi$ of the joint.

Expression (5.26) contains the *passive* elastic forces $F^{PE}(\phi)$ and $F_i^{PS} \sin \theta(\ell_i(\phi), \lambda_{Ti})$, where (see (5.20))

$$F_i^{PS} \triangleq - \bar{F}_i \, b_{1i} \, [2 - \exp\{6.97(\xi_i - 1)\}]. \quad (5.27)$$

In static isometric measurements these passive force contributions can always be eliminated from the measurement procedure, so that only the contractive contributions $F_i^{CE} \sin \theta(\ell_i(\phi), \lambda_{Ti})$ appear in the records. If this is done, the *predicted* total maximum *contractive* torque output follows from (4.54), (4.55), and (5.20) as

$$\tilde{T}_c = \sum_{i=1}^{m} D_i(\phi) \bar{F}_i \exp\{-(\frac{\xi_i - 1}{s_{ki}})^2\} \sin \theta(\ell_i(\phi), \lambda_{Ti})/(1-q_{oi}), \quad (5.28)$$

where the tilde on $\tilde{T}_c$ indicates that value of $T_c$ which corresponds to maximum stimulation.

81

On the other hand, a sequence of experimental maximum-effort torque measurements $\tilde{\mathcal{N}}(\phi_j)$, $j=1,\ldots,p$, can be made on the real biosystem at sequential angular positions $\phi_j$. We can then pose the constrained, weighted least-square parameter estimation problem of finding that parameter set

$$\{\bar{F}_i, s_{ki}, \bar{\alpha}_i, \bar{\ell}_i; i=1,\ldots,m; s_{ki} \geq 0.28\}, \quad (5.29)$$

which minimizes the discrepancy between the experimentally observed torque sequence $\tilde{\mathcal{N}}(\phi_j)$ and the predicted model torque sequence $\bar{T}_{cj}$, i.e. we minimize

$$\sum_{j=1}^{p} w_j \, [\tilde{\mathcal{N}}(\phi_j) - \bar{T}_c(\phi_j, \bar{F}_i, s_{ki}, \bar{\alpha}_i, \bar{\ell}_i)]^2 \quad (5.30)$$

with respect to the set (5.29). The symbols $w_j$, $j=1,\ldots,p$, denote weights (generally $w_j=1$).

The least-squares estimator is the criterion of choice since it provides good parameter estimates even for comparatively inaccurate data sequences, which is not necessarily true for other estimators (such as the min-max error criterion).

In order to perform the minimization of (5.30), the following functions and parameter values must be known: $D_i(\phi)$, $\ell_i(\phi)$, $q_{oi}$, $d_i$, $\theta_{oi}$, $\bar{\theta}_i$, $\lambda_{Toi}$, $b_{li}$, $\bar{\lambda}_i$, $i=1,\ldots m$, and $\lambda_{soi}$. Methods for the determination of $D_i(\phi)$, $L_i(\phi)$ (which is related to $\ell_i(\phi)$ via (5.25)), and $\lambda_{Toi}$ have been discussed in Section 5.1. For human muscle we always have $q_{oi} = 0.005$, which value then also determines the value of $b_{li} = q_{oi}/(1-q_{oi})$. Since the values of $\hat{\theta}_{oi}, \hat{\theta}_{oi}$, and $k_{si}$ are known (see Section 5.1), relations (4.47), (4.49) and (4.53) can be used to find the values of $\theta_{oi}, \bar{\theta}_i$, and $d_i$, where, however, the computation of $\bar{\theta}_i$ and $d_i$ involves values of the parameters to be estimated, and hence must be included in the iterative minimization procedure. Finally, $\bar{\lambda}_i$ and $\lambda_{soi}$ are given by (4.51) and (4.52) respectively. As mentioned above, in most cases involving penniform muscles it is possible to put $\lambda_{soi} \equiv 0$. For fusiform muscles, the value of $\lambda_{soi}$ is calculated from (5.6), while $\bar{\lambda}_i$ is given by (5.7). In general, the expressions for fusiform muscles follow as a special case from those for penniform muscles if we put $\theta \equiv \pi/2$ and $\lambda_s \equiv 0$, since the tendinous part of a fusiform muscle is identical with $\lambda_T$ of a penniform muscle (see Fig. 5.3).

The *computational procedure* for estimating the values of the parameter set (5.29) can now be summarized as follows.

(1) Provide initial estimates for the parameter values;
(2) minimize (5.30), where for each i, $i=1,\ldots,m$, and each j, $j=1,\ldots,p$, the value of $\lambda_{Ti}(\ell_i(\phi_j))$ is computed by solving (5.24);

stop the iterative minimization algorithm if the prescribed error bound is satisfied.

In Appendix A3 we describe and list the ANSI FORTRAN IV computer program ELPEST which executes the algorithm described above, i.e. for given functions $D_i(\phi)$ and $L_i(\phi)$, given values of the morphometric parameter set, and a given sequence $\tilde{\mathcal{N}}(\phi_j)$, $j=1,\ldots,p$, of experimentally observed maximum steady-state isometric torque outputs, the program computes least-squares estimates of the values of the parameter set (5.29), for a maximum of $m=10$ simultaneously acting penniform or fusiform muscles.

After having formulated the computational procedure required to obtain parameter estimates, we shall now give a detailed description of the *experimental procedure* to be followed in acquiring the torque sequence $\tilde{\mathcal{N}}(\phi_j)$, $j=1,\ldots,p$, for a given group. The procedure will be demonstrated for the human triceps muscle, and results of an actual experiment will be presented.

Subjects are seated, and strapped onto a special adjustable myodynamometer which makes it possible to observe isometric, concentric and eccentric contractions over a large range of angular positions and velocities. The output torque is measured by a highly sensitive and temperature-compensated torque strain gauge bridge which is mounted on the axis of the dynamometer. The whole arrangement is shown in Fig. 5.4.

In the present arrangement the output of the torque strain gauge bridge is amplified and fed into a vertical channel of a 9-channel Tektronix 5115 bistable storage oscilloscope. The horizontal channel is supplied with the output of a linear potentiometer which indicates the angular position $\phi$ of the forearm relative to the upper arm. Full extension of the forearm corresponds to $\phi = 0$ rad (position 1). Before each measurement, the subject is asked to relax, and the torque input to the oscilloscope is set to zero in order to eliminate all passive torques due to muscle structures and the weight of the limb.

At a given position $\phi_j$, the subject is instructed to produce a non-jerky isometric maximum-effort extension of the right forearm against the hand mould mounted on the dynamometer lever arm (Fig. 5.4). At each angle $\phi_j$, $j=1,\ldots,24$, the alignment of the approximate elbow centre with the centre of the dynamometer lever arm is checked and, when necessary, readjusted. Sufficient resting time must be allowed between the contractions, which are to be performed in a random sequence at 24 angular positions $\phi_j, j=1,\ldots,24$. The positions are chosen by selecting the first two digits from a 4-digit random number table. In our case the sequence $\{j\}$ was: 4, 16, 12, 18, 2, 7, 5, 23, 24, 9, 20, 19, 1, 22, 14, 3, 21, 15, 6, 11, 17, 10, 8,13. This procedure guarantees the

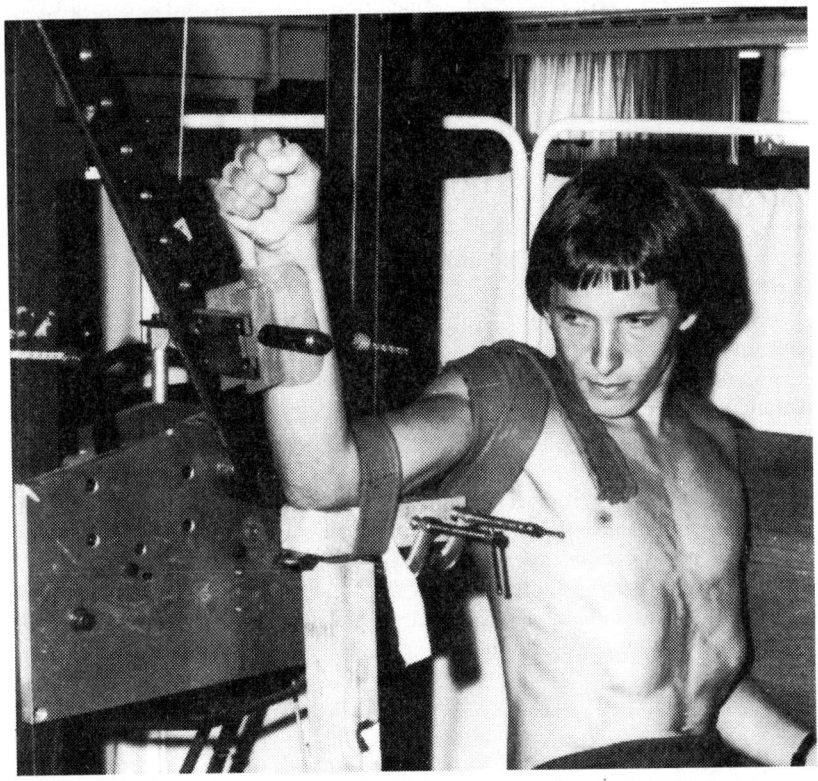

**FIG. 5.4:** Experimental arrangement for the measurement of arm extensor torques

elimination from the experimental records of any bias due to fatigue or training effects. Calibrations of both output torque and angular position are checked at the beginning and end of an experimental series. The increment $\triangle \phi$ between successive angular positions $\phi_j$ and $\phi_{j+1}$ was 0.108 rad. Each contraction produces a vertical bar on the oscilloscope. These bars are stored, and after completion of the whole measuring sequence a photograph is taken of the screen. A typical such record is shown in Fig. 5.5.

At the end of each experimental session the forearm length $\ell_{fa}$, upper arm length $\ell_a$, and maximum circumference $u_a$ of the upper arm are measured on each subject. The dimensions $\ell_{fa}$ and $\ell_a$ are, of course, needed in (5.15), (5.16), and (5.17), while $u_a$ is required to enable approximate initial estimates of $\bar{F}_i$, $i=1,2,3$, to be computed according to the relations (details in Hatze, 1980a).

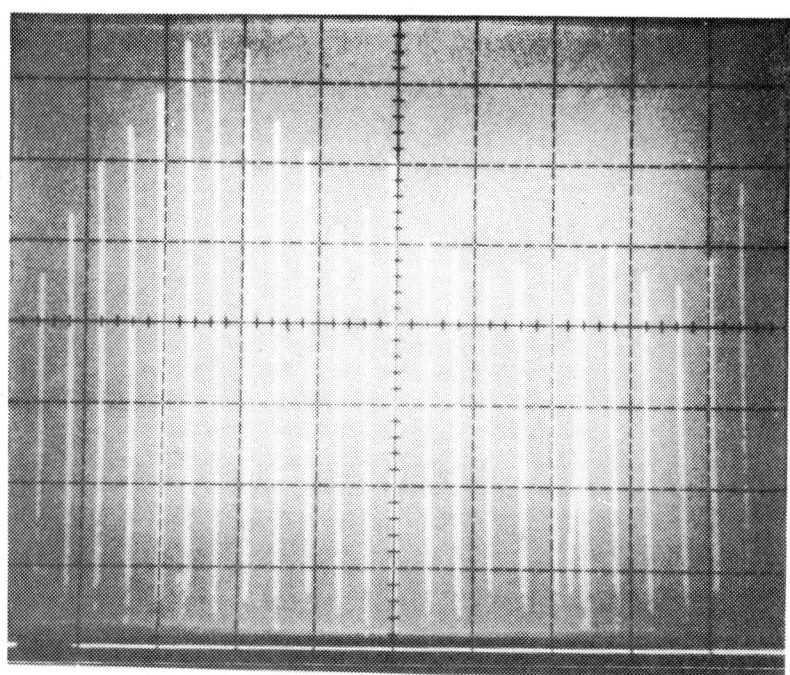

**FIG. 5.5:** Oscilloscope photograph of a sequence of isometric maximum-effort torque outputs of the triceps muscle. Calibrations: vertical scale: 11.6 Nm/div; horizontal scale: 0.25 rad/div

$$\bar{F}_1 \approx 925000 \, A_{Tr}^a, \quad \bar{F}_2 \approx 443000 \, A_{Tr}^a, \quad \bar{F}_3 \approx 425000 \, A_{Tr}^a,$$

where

$$A_{Tr}^a = 0.0369 \, u_a^2 - 0.0069 \, \ell_a^2,$$

the units being N and m. $\bar{F}_1$ refers to the medial head, $\bar{F}_2$ to the lateral head, and $\bar{F}_3$ to the long head of the triceps muscle. The estimates are based on an absolute force of 100 N per cm² of physiological cross-sectional area.

The experimental and computational procedures described above were applied to a number of male and female subjects. Fig. 5.6 shows the results for a case in which the torque sequence $\tilde{\mathcal{N}}(\phi_j), j=1,\ldots,22$, is subject to a comparatively large measurement error. As can be seen, the method still produces good estimates.

The computed estimates of the values of the parameter set (5.29), which correspond to the data depicted in Fig. 5.6, as well as associated

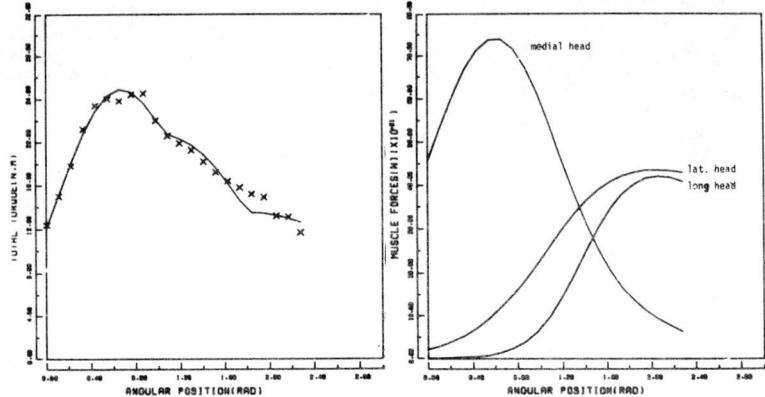

**FIG. 5.6:** *Left-hand diagram:* Comparison of measured torque (crosses) and least-squares fitted model torque (line) for a female subject aged 23. *Right-hand diagram:* Corresponding isometric length-tension curves of the three heads of the triceps muscle. Note that the curves are here presented as functions of the joint angle $\phi$.

constants are listed in Table 5.3. The values of the morphometric constants used in the computations are $\ell_{fa} = 0.239$ m, $\ell_a = 0.243$ m, and $u_a = 0.209$ m.

**TABLE 5.3:** Computed values of parameters and associated constants corresponding to the data shown in Fig. 5.6. Units are N, m, and rad

| Triceps muscle | $\bar{F}_i$ | $s_{ki}$ | $\bar{\alpha}_i$ | $\bar{\ell}_i$ | $\bar{\lambda}_i$ | $d_i$ | $\bar{\theta}_i$ |
|---|---|---|---|---|---|---|---|
| medial head (i=1) | 805.4 | 0.433 | 0.076 | 0.086 | 0.055 | 0.017 | 1.275 |
| lateral head (i=2) | 482.4 | 0.293 | 0.072 | 0.141 | 0.105 | 0.031 | 1.279 |
| long head (i=3) | 486.0 | 0.284 | 0.071 | 0.089 | 0.059 | 0.018 | 1.276 |

A more detailed discussion of these results can be found in (Hatze, 1980a).

### 5.2.2 Observations of dynamic isometric torque outputs at a fixed muscle length

The parameters whose values are still to be estimated are given by the set

$$\{\bar{c}, \hat{F}^C/\bar{F}, a_2', a_3', \check{\xi}\}. \tag{5.31}$$

Since *maximum steady-state* isometric contractions do not contain dynamic phases, parameters which characterize these phases are missing from the equations describing this contractive mode, and hence these parameters cannot be determined from torque observations made under steady-state conditions. It is, however, possible to utilize experimental observations of *dynamic* (i.e. time-changing) isometric torque outputs for the estimation of the values of the set (5.31). To this end, let it be assumed that a group of m' *penniform muscles,* all of which exhibit different myodynamic characteristics, contribute collectively to the externally observed total dynamic torque output $\tilde{\mathcal{N}}(t_j)$, $j=1,\ldots,p$. It will also be assumed that the muscle group is kept at a *fixed length* $\ell^*$. If we consider only the rising phase of an isometric contraction starting from rest, the differential system (4.37) reduces to

$$\dot{n}_i = \hat{n}_i z_i(t), \qquad\qquad n_i(0) = 0,$$

$$\dot{\psi}_i = m(n_i)\,[cv_i(t)-\psi_i] + z_i(t)\bar{c}_i\hat{n}_i \frac{1-\exp\{\rho_o(\xi_i)\psi_i\}}{\rho_o(\xi_i)\,(1-\exp\{-\bar{c}_i n_i \cdot \bar{\delta}\})}, \quad \psi_i(0) = 0,$$

(5.32)

$$\dot{\xi}_i = \frac{1}{S_i}\,[\frac{1}{a_{3i}}\ \text{arcsinh}\ \{-\frac{1}{a_{2i}}\ell n\{\frac{k(\xi_i)\epsilon_i}{b_{2i}[F_i^{SE}/\bar{F}_i + b_{1i}k_i(\xi_i)]} - a_{1i}\}\}-\frac{1}{2}],$$

$$\xi_i(0) = \xi_{oi},$$

for each of the $i=1,\ldots,m'$ penniform muscles, where $F_i^{SE}/\bar{F}_i$ is given by

$$F_i^{SE}/\bar{F}_i = [\exp\{\frac{\sigma}{\bar{\alpha}_i \lambda_{soi}}\ ([d_i^2+(\ell_i^*-\lambda_{Ti})^2]^{1/2}-\bar{\lambda}_i(\xi_i+\bar{\kappa})-\lambda_{soi})\}-1]/[\exp\{\sigma\}-1],$$

(5.33)

where we may put $\bar{\kappa} \equiv 0$ in accordance with the discussion in Chapter 4.

It will be assumed that all passive torques (due to passive muscle structures, limb weight, etc.) have been eliminated from the experimental records, and that $\ell^*$ has been so chosen that the resting force contributions $F_{oi}^{SE}/\bar{F}_i$, $i=1,\ldots,m'$, become negligible. If possible, $\ell^*$ should be chosen to be near the weighted mean optimum length

$$\bar{\ell} = \sum_{i=1}^{m'} \bar{F}_i \bar{\ell}_i\ /\ \sum_{i=1}^{m'} \bar{F}_i. \qquad (5.34)$$

Under these conditions only the *purely contractive force contributions* $F_i^{CE} \sin\theta(\ell_i^*, \lambda_{Ti})$ appear in the records, and the *predicted* total *contractive* torque output follows from (4.55) as

$$T_c = \sum_{i=1}^{m'} D_i(\phi^*)\bar{F}_i \sin\theta(\ell_i^*, \lambda_{Ti})[F_i^{SE}/\bar{F}_i + b_{1i}k_1(\xi_i)], \quad (5.35)$$

since the purely contractive contribution $F_i^{CE}/\bar{F}_i$ is defined by

$$F_i^{CE}/\bar{F}_i = F_i^{SE}/\bar{F}_i + b_{1i}k_1(\xi_i).$$

The unknown $\lambda_{Ti} = \lambda_{Ti}(t_j)$ can be found by solving, at each sampling time $t_j$, $j=1,\ldots,p$, the nonlinear equation $F_i^{ST} = F_i^{SE}\sin\theta(\ell_i^*,\lambda_{Ti})$, i.e.

$$\frac{\sin\bar{\theta}_i}{\sin\theta(\ell_i^*,\lambda_{Ti})}[\exp\{\sigma(\lambda_{Ti}/\lambda_{Toi}-1)/\bar{\alpha}_i\}-1]/[\exp\{\sigma\}-1] = F_i^{SE}(\ell_i^*,\lambda_{Ti},\xi_i)/\bar{F}_i$$

$$(5.36)$$

where $F_i^{SE}(\cdot)/\bar{F}_i$ is given by (5.33) with $\bar{\kappa} \equiv 0$.

In many cases we can assume that the tendinous fibre parts of a penniform muscle are short compared with the length of the tendon proper, i.e. we may put $\lambda_{soi} = 0$. Under these circumstances (5.35) becomes

$$T_c = \sum_{i=1}^{m'} D_i(\phi^*)\bar{F}_i [F_i^{ST}/\bar{F}_i + b_{1i}k_1(\xi_i)\sin\theta(\xi_i)], \quad (5.37)$$

where $F_i^{ST}$ is given by (4.54) with $\lambda_{Ti} = \lambda_{Ti}(t_j)$ expressed by

$$\lambda_{Ti}(t_j) = \ell_i^* - [(\bar{\lambda}_i\xi_i(t_j))^2 - d_i^2]^{1/2}. \quad (5.38)$$

Furthermore,

$$\sin\theta(\xi_i(tt_j)) = [1-(d_i/\bar{\lambda}_i\xi_i(t_j))^2]^{1/2},$$

$\xi_i(t_j)$, $n_i = n_i(t_j)$ and $\psi_i = \psi_i(t_j)$ are defined by (5.32), and $F_i^{SE}/\bar{F}_i$ in (5.32) must be replaced by $F_i^{ST}/(\bar{F}_i\sin\theta(\xi_i))$. The model muscle lengths $\ell_i^*$ are given by

$$\ell_i^* = \ell_i' + \Delta L(\phi^*), \quad i=1,\ldots,m', \quad (5.39)$$

where for penniform muscles $\ell_i'$ is related to the shortest real muscle length $L_i'$ by (5.25). Note that no nonlinear equation has to be solved for the case where $\lambda_{soi} = 0$.

In the case of *fusiform muscles*, Equation (5.33) applies, with $[d_i^2 + (\ell_i^* - \lambda_{Ti})^2]^{1/2}$ replaced by $\ell_i^*$, and $\bar{\kappa} \equiv 0$. The total purely contractive torque output $T_c$ is then given by (5.35) with $\sin\theta(\ell_i^*,\lambda_{Ti}) \equiv 1$.

Under maximum neural stimulation ($z_i(t) = 1$, $v_i(t) = 1$, $i=1,\ldots,m'$) and after a sufficiently long time (about one second), $T_c(t)$ attains its isometric steady-state maximum value $\tilde{T}_c$. The corresponding experimentally observed torque output is $\mathcal{N}(\phi^*)$. Hence (5.35) and (5.37) can be normalized by dividing through by the predicted isometric maximum torque $\tilde{T}_c$, while the experimentally observed torque sequence $\mathcal{N}(\phi^*,t_j)$, $j=1,\ldots,p$, is normalized by dividing through by $\mathcal{N}(\phi^*)$.

In analogy to (5.30) we can pose the weighted least-square parameter estimation problem of finding that parameter set

$$\{\bar{c}_i, \hat{F}_i^C/\bar{F}_i, a'_{2i}, a'_{3i}, \check{\xi}_i;\ i=1,\ldots,m'\} \tag{5.40}$$

which minimizes

$$\sum_{j=1}^{p} w_j\, [\mathcal{N}(\phi^*,t_j)/\mathcal{N}(\phi^*) - T_c(\phi^*,t_j,\ \bar{c}_i, \hat{F}_i^C/\bar{F}_i, a'_{2i}, a'_{3i}, \check{\xi}_i)/\tilde{T}_c]^2, \tag{5.41}$$

with respect to the set (5.40), and where generally $w_j = 1$, $j=1,\ldots,p$.

The advantage of the normalization is that for the frequently occurring case where $D_i(\phi^*) = D(\phi^*)$, $i=1,\ldots,m'$, the constant $D(\phi^*)$ cancels in $T_c/\tilde{T}_c$, and hence need not be known.

The problem of minimizing (5.41) is considerable more difficult than that of minimizing (5.30).

First, for each estimate of the values of the parameter set (5.40), the differential system (5.32) must be integrated numerically to yield the values $n_i(t_j)$, $\psi_i(t_j)$, and $\xi_i(t_j)$, $j=1,\ldots,m'$, which are required in (5.35) or (5.37) for the computation of $T_c(t_j)$.

Secondly, the neural control input functions $z_i(t)$ and $v_i(t)$, $i=1,\ldots,m'$, $t \in [t_1,t_p]$, must be estimated, by some method, from electromyographic observations recorded simultaneously with the corresponding output torques $\mathcal{N}(\phi^*,t)$. The normalized stimulation rates $v_i(t)$ can be computed from (4.39), provided estimates of the motor unit stimulation rates $v(x,t) = 1/\tau_i(x,t)$ (see (4.40)) are available. However, such estimates are difficult to obtain. At present, only multisample needle or indwelling-wire electromyography can provide the necessary information and then only for comparatively weak contractions. In addition, these estimates are required for each of the $m'(m' \leq 10)$ simultaneously contracting muscles, a fact that further complicates the situation.

Finally, *in situ* estimates of values of the maximum relative stretching force $\hat{F}^C/\bar{F}$ of a muscle are also difficult to obtain. The reason is that during fast eccentric contractions, stretch reflexes are activated which make it difficult to separate the force component due to reflex actions from that which is due to the force-velocity properties only. However, animal experiments (Joyce and Rack, 1969) and experiments performed on human muscle (Komi, 1973) indicate that the value of $\hat{F}^C/\bar{F}$ for mammalian skeletal muscle is about 1.30–1.33. For $\hat{F}^C/\bar{F} = 1.33$, the corresponding force-velocity constants are given in Table 5.1.

A detailed description of the experimental and computational procedures for estimating the neural input control functions $z_i(t)$ and $v_i(x,t), i=1,\ldots,m'$, from electromyographic recordings is far beyond the scope of this monograph. Indeed, these models and techniques are so complicated that their detailed description requires a separate treatise. For this reason, the present description of the *experimental procedures* required for estimating the values of the parameter set (5.40) must, by necessity, be incomplete, and will therefore be confined to an outline of the procedures to be followed.

We shall now discuss two experimental methods which can be used to estimate the values of the set (5.40), excluding $\hat{F}_i^C/\bar{F}_i$.

### 5.2.2.1 Observations of quasi-stationary torque outputs at various activation levels

In this contractive mode the subject is required to maintain successive constant levels of prescribed submaximal isometric force (torque) outputs. Simultaneous electromyographic needle recordings make it possible to estimate the relative numbers $n_i(T_{cj})$, $i=1,\ldots,m'$, of active motor units and the individual motor unit firing rates $v_i(x,T_{cj})$ of the $i=1,\ldots,m'$ muscles, at a prescribed relative output torque level $T_{cj}/\tilde{T}_c$. With $v_i(x,T_{cj}) = 1/\tau_i(x,T_{cj})$ known, $v_i(x,T_{cj})$ can be computed from (4.40), which in turn can be used to obtain $\bar{v}_i(T_{cj})$ from (4.39). Since for this contractive mode $z_i(T_{cj}) \equiv 0$, $\dot{\xi}_i \equiv 0$, and $\dot{\psi}_i \equiv 0$, the latter implies that

$$\psi_i(T_{cj}) = c\, v_i(T_{cj}), \qquad (5.42)$$

while $\dot{\xi}_i \equiv 0$ implies,

$$(1-q_{oi})\, [F_i^{SE}/\bar{F}_i + b_{1i}k_1(\xi_i)] = k(\xi_i)\varepsilon_i. \qquad (5.43)$$

Equation (5.43) is a nonlinear equation in $\xi_i$ which must be solved together with (5.33) for fusiform muscles, and together with (5.33) and (5.36) for penniform muscles (to obtain $\lambda_{Ti}$). The predicted contractive torque $T_{cj}$ can then be found from (5.35) or (5.37), while $\tilde{T}_c$ is obtained by putting $n_i = 1$, $v_i = 1$, $i=1,\ldots,m'$.

The advantage of this scheme is that no integration of (5.32) is required, while it has the disadvantage that it yields only rather imprecise estimates of the values of $a'_{2i}$ and $a'_{3i}$, $i=1,\ldots,m'$, which parameters appear only in equation (4.40) for this contractive mode. For this reason, this contractive mode is not recommended for the present purpose.

### 5.2.2.2 Observations of linearly increasing torque outputs

In this contractive mode the subject is required to follow approximately a prescribed, linearly increasing torque. The joint concerned is fixed (isometric condition) at an angle $\phi^*$ corresponding to a muscle length $\bar{\ell}$ as defined by (5.34). Contractions start with the muscle at rest and proceed until a submaximal, or even maximal level has been reached. At the beginning of the contractions, all passive torque contributions (due to passive tissue structures, limb mass, etc.) must have been eliminated from the torque output record. This can usually be done electronically by removing the bias output voltage of the strain gauge amplifier.

The prescribed rate of linear torque increase can be adjusted by the experimenter for successive runs. This enables the experimenter to select the most suitable record for analysis. Appropriate values for the rate $d(T_c(t)/\bar{T}_c)/dt$ of prescribed relative torque increase have been found to be in the range $0.05 - 1$ s$^{-1}$, i.e. the maximum isometric muscle torque is reached from rest in times ranging between 20 s and 1 s.

One possible way of simultaneously displaying to the subject the prescribed torque function and the torque produced by the muscle group, is to connect the output of a triangular-wave generator to one vertical input channel of a dual-beam oscilloscope, and the output of the myodynamometer strain gauges to a second vertical input channel. The horizontal axis is the common time axis. In this way it is possible for the subject to track the prescribed ramp torque by adjusting his own force production so as to approximately match the prescribed value. It must, however, be emphasized that it is *not* essential for the subject to track the prescribed torque precisely. The prescription of a linearly increasing output is merely a convenient means of exercising some external control over the contractive mode. In the final analysis, the neural controls $z_i(t)$ and $v_i(t)$, $i=1,\ldots,m'$, as extracted from electromyographic observations, are in any case correlated with the observed relative output torque values $\mathcal{N}(\phi^*,t)/\bar{\mathcal{N}}(\phi^*)$. Hence the association of particular input functions $z_i(t)$, $v_i(t)$, with the corresponding torque recordings $\mathcal{N}(\phi^*,t)$, guarantees the independence of the parameter identification process from any specific dynamic contractive mode.

Finally, in order that the value of $\bar{\mathcal{N}}(\phi^*)$ which is needed in (5.41) may be determined, a maximum-effort contraction has to be performed by the subject.

Fig. 5.7 shows the experimental arrangement used in the present procedure, while Fig.5.8 displays some of the electromyographic and torque records of a ramp contraction performed by the quadriceps femoris muscle group at a rate of 0.5 s$^{-1}$.

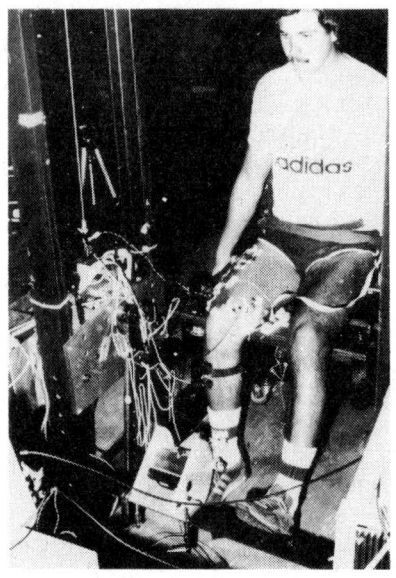

FIG. 5.7: Experimental arrangement for the recording of the isometric output torque, and the surface and indwelling-wire electromyograms of the muscles of the quadriceps femoris group. Two wire electrodes (at different locations) and one surface electrode are used on the rectus femoris, vastus lateralis, and vastus medialis to record the electromyograms of each of these muscles

Upon completion of the experiment, the electromyographic records are analyzed by means of special methods (too elaborate to be discussed here), yielding the control functions $z_i(t)$ and $v_i(t)$, $i=1,\ldots,m'$, needed for the integration of (5.32). The values of the myocybernetic constants $\hat{n}_i$, $i=1,\ldots,m'$, which are also required in (5.32), are estimated from electromyographic recordings of maximum isometric contractions (see Section 5.3). The observed torque function $\mathcal{N}(\phi^*,t)$ is now sampled at discrete time points $t_j$, $j=1,\ldots,p$, $p \leq 12m'$, and divided by the maximum torque $\tilde{\mathcal{N}}(\phi^*)$. The required set of parameter values is then found by minimizing (5.41).

It is important that good initial estimates of the parameter values should be supplied, because the starting point of the algorithm is of great relevance, since multiple minima may occur.

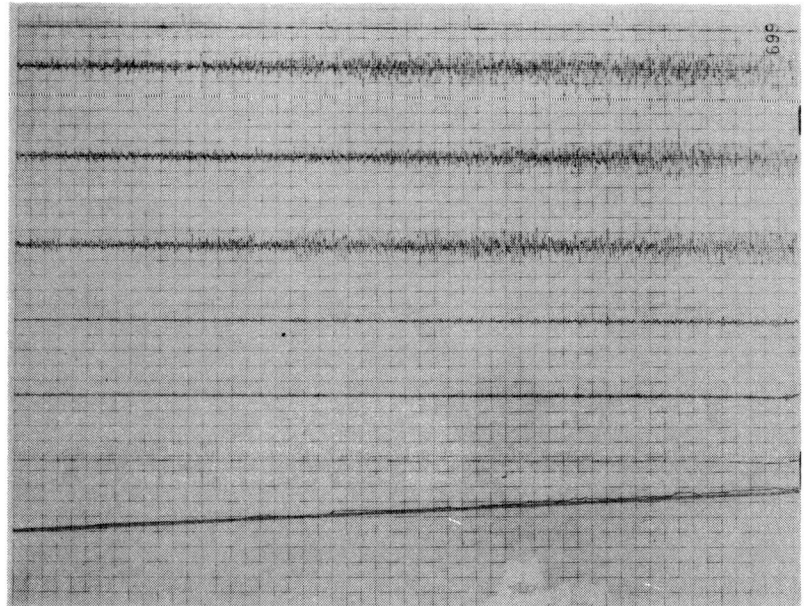

**FIG. 5.8:** Electromyographic (wire electrodes) and output torque recordings of an isometric ramp (i.e. linearly increasing) contraction of the quadriceps femoris muscle group. Only three out of a total of nine EMG records stored on a tape recorder are shown here. The traces from top to bottom are: time marker (1 second), EMGs of vastus medialis, rectus femoris, vastus lateralis, hamstrings, gastrocnemius, and prescribed ramp function (thick line) tracked by the muscle torque (thin line). Arbitrary scales. The light patches in the torque record are due to the deletion of notes made by the experimenter

## 5.3 MYOCYBERNETIC METHODS

The set of myocybernetic parameters whose values must still be estimated is given by (5.9). The parameter $\hat{n}$ designates the maximum rate of *motor unit recruitment,* while the product $\hat{n}\,\check{z}$, defining the parameter $\check{z}$, denotes the maximum rate of *motor unit derecruitment* of a given muscle.

The rate $\hat{n}$ is the inverse of the time required for all the motor units of a muscle to be recruited in a *maximum-effort* contraction, starting from rest. As mentioned above, electromyographic techniques (not to be discussed in this context) can be used to estimate the value of $\hat{n}$. For medium-size skeletal muscles (such as the rectus femoris) a fairly representative average value for $\hat{n}$ is 20 (50 ms minimum recruitment time), while the corresponding value for large muscles is about 14.2 (70

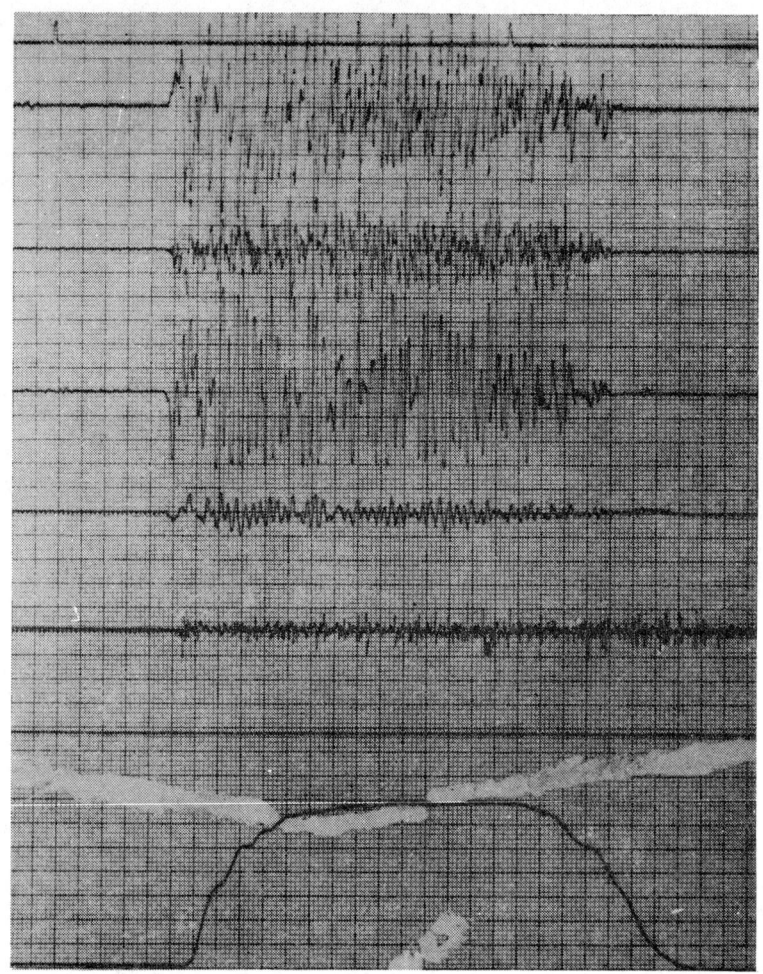

**FIG. 5.9:** Electromyographic (surface electrodes) and output torque recordings of a maximum-effort, recruitment-derecruitment contraction of the quadriceps femoris muscle group. The traces from top to bottom are: time marker (1 s), EMGs of vastus medialis, rectus femoris, vastus lateralis, hamstrings (ten times scale of previous recordings), gastrocnemius (ten times scale of vastus lateralis), and output torque (39.8 Nm/div). Arbitrary scales for EMGs. Note the late burst of EMG activity during relaxation, elicited by a stretch reflex. The ligth patches in the torque record are caused by the deletion of spurious traces

ms min. recr. time). The electromyograms, from which estimates of $\hat{n}$ and $\check{z}$ are obtained, and the force output of the maximally contracting muscles of the quadriceps femoris group of a male subject aged 24 are shown in Fig. 5.9.

It must be emphasized that the assumption of a constant rate $\hat{n}$ of motor unit recruitment is based on experimental evidence presently available. This evidence suggests that in maximum-effort contractions the recruitment intervals are approximately constant and hence independent of the motor unit size. It is, however, possible that more refined experimental techniques could reveal a dependence of $\hat{n}$ on n, conceivably of the form $\hat{n} = \hat{n}^*/(1-a_3'n/a_2')$, where $\hat{n}^*$ denotes a new constant.

The value of the derecruitment parameter ž is defined by

$$\check{z} = 1/\hat{n}\check{t}, \tag{5.44}$$

where ť is the time required for all the motor units of a muscle to be derecruited at maximum rate, starting from full activation. Again, electromyographic methods are used to estimate the value of this parameter. It turns out that in general the minimal derecruitment times are about the same as the minimal recruitment times, as is clearly illustrated in Fig. 5.9. Hence $\check{z} \approx 1$.

This completes the discussion of the methods used for estimating myodynamic and myocybernetic parameter values. It should perhaps be mentioned that the values of certain parameters relating to the properties of the *inactive (passive)* muscle must be determined by techinques which are fundamentally different from those described here. In fact, in most cases it is impossible to separate the contributions made by the muscle structures to the passive visco-elastic joint torques, from those originating from other structures such as ligaments, connective tissue, etc. For this reason, these methods will not be discussed in the present context.

## CHAPTER 6

# Model validation and computer experiments

The *credibility* of a model is determined by its verifiable predictions. It is therefore imperative to test under as wide a variety of conditions as possible the various myocybernetic models of skeletal muscle developed in the previous chapters. Since the models are essentially global in nature, the *domain of applicability* includes all control and contractive modes consistent with macro-myodynamics.

In order to simulate on the computer the responses of the mathematical models to a variety of control inputs, a *computer program* implementing these conceptual models must be developed. In Appendix A4 the complete listing is given of the computer program MYOSIM simulating the responses of four different myocybernetic models. These computerized models have been *verified,* i.e. it has been established that they represent the conceptual models adequately, within specified limits of accuracy.

In this chapter we shall be concerned mainly with the *validation* of the computerized models, i.e. with the process of establishing their ability to predict system responses that agree, within specified limits of accuracy, with the corresponding responses as observed on the real biosystem (i.e. the myostructures). In addition, we shall infer some theoretical results from the simulation studies, and perform computer experiments on the various muscle models.

It should be noted that the simulations performed by the computer program MYOSIM can be used most advantageously for teaching purposes. The program has a graphic-display option which makes it possible for the user to display on the electronic screen of a graphics terminal, or on a paper plot, all dynamic responses of the various models. The user can thus perform a large variety of 'computer experiments' on the muscle models, including isometric, eccentric, and concentric modes of contraction; stimulation at time-varying rates of single muscle fibres, groups of fibres, single motor units, groups of motor units, and a whole muscle consisting of a large number of different-sized motor units; and he can immediately observe the response of the model muscle in the form of a graphic display. Numerous examples of such displays will be given below. Four models are available for this purpose.

*Model 1* simulates the dynamic behaviour of N(maximally 10) simultaneously acting *single fibres* of a fusiform muscle. The model equations are given by (3.21), (3.22), (3.23), (3.29) (with $\rho(\xi) \equiv 66200$ since the length-dependence of $\gamma$ has already been accounted for in (3.21)), (3.45), and (3.7), where the neural controls are individual trains of nerve impulses $\alpha_i(t)$, $i=1,\ldots,N$, as defined by (3.23). The stimulation rate $\tau_i^{-1}(t)$ for each of the N muscle fibres can be specified by the experimenter. The simulation responses observable for each of the $i=1,\ldots,N$ muscle fibres, and for the whole time interval [TSTART, TFIN] of the simulation are: the individual (normalized) action potentials $\beta_i(t)$; the (normalized) interfilamentary calcium concentrations $\gamma_i(t)$; the normalized contractile element lengths $\xi_i(t)$; the active states $q_i(t)$; and the normalized fibre force outputs $f_i^{SE}(t)/\bar{f}_i$. In addition, the total normalized force output of all N fibres and the muscle length $\ell(t)$ can also be observed. The experimenter can prescribe the length function $\ell(t)$ so as to induce stretches, shortenings, or isometric conditions.

*Model 2* – similar to MODEL 1, except that the *excitation dynamics* is now given by (3.27), replacing Equations (3.21)–(3.23).

*Model 3* simulates the control behaviour of a *whole fusiform muscle* consisting of a large number of motor units, all of which differ in size and type. The model equations are given by (4.37), (4.41), (4.42), and other equations associated with these. The neural control inputs $z(t)$ and $v(t)$, as well as the length function $\ell(t)$, can again be specified by the experimenter. The *observable model responses are:* the normalized population $n(t)$ of stimulated motor units; the normalized population $r(t)$ of semi-active motor units; the normalized length $\xi(t)$ of the contractile element; the active states $q_n(t)$ and $q_r(t)$; the normalized total force output $F(t)/\bar{F}$; and the muscle length $\ell(t)$.

*Model 4* – similar to MODEL 1, except that single muscle fibres are replaced by single *motor units* of exponentially increasing sizes, according to (4.18) and (4.19). The control and contractive modes may be selected by the experimenter as in MODEL 1.

A very detailed description of how the computer program is to be used is given in the introductory comments of the program listing in Appendix A4. Experience has shown that such computer experiments on appropriate models of biosystems constitute an invaluable teaching and research aid for students of the physiological, bioengineering, medical, physical, and sports sciences.

## 6.1 COMPUTERIZED MYOCYBERNETIC MODELS

To implement on the digital computer the mathematical (conceptual) models of skeletal muscle discussed above, it is necessary to take into account the limitations imposed on the inherently continuous model dynamics by the digital structure of the computing system. The differential systems describing the dynamic behaviour of the models must be integrated numerically in finite, discrete steps, and subject to truncation and roundoff errors. This can, and does, lead to numerical problems in the simulation of complex and highly nonlinear systems such as these. However, these problems can be overcome by applying appropriate numerical techniques, and by introducing certain minor modifications that do not alter the conceptual bases of the models.

A major problem in simulating the contraction dynamics (3.45) or the last of Equations (4.37) was the occurrence of negative values of the argument in the logarithmic function. This is due to the fact that $f^c(\dot{\eta},.)$, as given by (3.40), is defined for all $\dot{\eta}$ with values on the real line, while the inverse function $\dot{\eta}(f^c,.)$ (or equivalently $\dot{\xi}(f^c,.)$) as expressed by (3.45)) is defined only for those values of $f^c/\bar{f}$ that ensure a non-negative argument of the logarithmic function for any given set of values $\{a_1, b_1, b_2, a_6', \xi(t), \gamma(t)\}$. It can be shown that in the case of analog simulations $f^c/\bar{f}$ always remains within its domain of definition provided the simulation commences from a permissible initial state. For digital simulations, however, the situation is different.

If fast stretches are applied to the model muscle, it may happen that the discrete increment $\triangle \xi(t_1)$, corresponding to the discrete and finite integration interval $\triangle t$, is insufficient appropriately to offset the increment in $\delta(t_1)$ (see (3.7)) produced by the stretch $\triangle \ell(t_1)$. In this case $f^c/\bar{f} > \hat{f}^c/\bar{f}$, thus exceeding its domain of definition. This could, of course, never happen in a continuous simulation.

To overcome this problem we redefine the function $f^c(\dot{\eta}, q=1, \xi=1)/\bar{f}$, as given by (3.40), in such a way that for $\dot{\eta} > \dot{\eta}^*, \dot{\eta}^* > 0$, the original function is replaced by a linear function with a sufficiently small and positive gradient. This closely mimics physiological reality and at the same time eliminates the above problem.

In practice we proceed as follows. Instead of expressing $g(\dot{\lambda}) = g(\dot{\eta})$ by (3.35), we define the linear function

$$g(\dot{\eta}) \triangleq a_o + \bar{\omega}\bar{f}\dot{\eta}, \quad \dot{\eta} \geq \dot{\eta}^*, \quad (6.1)$$

where $\dot{\eta}^*$ is given below and the gradient $\bar{\omega}$ is a function of $\hat{f}^C/\bar{f}$, i.e.

$$\bar{\omega} \triangleq 0.01(1/a_1b_2-b_1-1), \qquad (6.2)$$

where (3.43) has been used.

In analogy to (3.39) we have for $\dot{\eta} \geq \dot{\eta}^*$,

$$f^C = q(\cdot)k(\xi)(a_o + \bar{\omega}\bar{f}\dot{\eta}) - b_1\bar{f}k_1(\xi). \qquad (6.3)$$

We now define $\dot{\eta}^*$ by the relation

$$f^C(\dot{\eta}^*, q=1, \xi=1) = 0.999\hat{f}^C = 0.999\bar{f}(1/a_1b_2-b_1). \qquad (6.4)$$

Substitution of (6.4) into (6.3) yields

$$a_o = \bar{f}(0.999/a_1b_2 + 0.001b_1 - \bar{\omega}\dot{\eta}^*), \qquad (6.5)$$

so that (6.3) becomes

$$f^C/\bar{f} = q(\cdot)k(\xi)[0.999/a_1b_2 + 0.001b_1 + \bar{\omega}(\dot{\eta}-\dot{\eta}^*)] - b_1k_1(\xi). \qquad (6.6)$$

Since $f^C = f^{SE}$, and by virtue of (3.37), we obtain from (6.6):

$$\dot{\xi} = \frac{-\bar{\lambda}_o}{\bar{\lambda}}[\dot{\eta}^* + \frac{1}{\bar{\omega}}\{\frac{f^{SE}/\bar{f} + b_1k_1(\xi)}{q(\xi,\gamma)k(\xi)} - \frac{0.999}{a_1b_2} - 0.001b_1\}], \qquad (6.7)$$

where $k_1(\xi) = 2 - \exp\{a_6'(\xi-1)\}$, and the constant $\dot{\eta}^*$ is defined by

$$\dot{\eta}^* = \frac{1}{a_3}\operatorname{arcsinh}\{-\frac{1}{a_2}\ell n\{\frac{1}{b_2[0.999(1/a_1b_2-b_1)+b_1]} - a_1\}\} - \tfrac{1}{2}. \qquad (6.8)$$

The expression $0.999(1/a_1b_2-b_1) = 0.999\hat{f}^C/\bar{f}$ denotes the critical force level for $\xi = 1$, $q = 1$. In general, if

$$f^{SE}/\bar{f} > 0.999[q(\xi,\gamma)k(\xi)/a_1b_2 - b_1k_1(\xi)], \qquad (6.9)$$

the original contraction dynamics as given by (3.45), or the last of Equations (4.37), is replaced by its linear extension (6.7).

A similar problem arises when the initial length $\ell_o$ of the resting muscle (fibre) is less than $(\bar{\ell} - \bar{\alpha}\lambda_{so})$ at the beginning of the simulation. In this case the tendinous part of the muscle (fibre) is slack and hence we must put $f^{SE}/\bar{f} = 0$ (or $F^{SE}/\bar{F} = 0$ in model (4.37)) because the tendons cannot support compressive forces. Then, if the initial state $\xi^o$ had been given a value smaller than unity, the expression $b_1(2-\exp\{a_6'(\xi^o-1)\})$ in (3.45) (or $b_1k_1(\xi^o)$ in (4.37)) would increase with decreasing values of $\xi^o$, while $k(\xi^o)$ would decrease, and $q = q_o$ (resting state) would remain constant. This could again lead to a negative argument of the logarithmic function in (3.45) or (4.37).

However, the allocation to $\xi^o$ of a value smaller than unity under the condition of a slack tendon is erroneous, since Gordon *et al.* (1966) and

Ramsey and Street (1940) have demonstrated experimentally that the resting muscle always extends itself to its rest length such that $\xi^o = 1$. Hence if $f^{SE}/\bar{f} = 0$ (or $F^{SE}/\bar{F} = 0$), $\xi^o < 1$, and the argument of the logarithmic function becomes negative, we put

$$\dot{\xi} = \max\{(1-\xi)/\triangle t, 0.01/\triangle t\}, \qquad (6.10)$$

which rapidly returns the contractile element to its equilibrium length $\xi^o = 1$. The symbol $\triangle t$ denotes the integration step size.

Other computational problems relating to the integration of the differential system (4.37) are discussed exhaustively in Appendix A1, and will therefore not be repeated here. The user of the computer program MYOSIM need, however, not be concerned with these details since the program automatically implements all the modifications discussed above.

We shall now proceed to the validation of the various myocybernetic models by performing computer experiments on the respective model muscles (fibres).

## 6.2 COMPUTER EXPERIMENTS ON MODEL MUSCLE FIBRES

In this section we simulate responses of single muscle fibres to a variety of control and contractive modes. We begin by simulating twitch responses.

### 6.2.1. Simulation of twitch responses

A twitch is defined to be the response of the myostructures to a single nerve input volley.

The first experiment is designed to test the hypothesis that the substitution in Equation (4.4) of the function $h(\dot{\xi}_{ik})$ for the function $H(\dot{\lambda}_{ik})$ is justified. Expressed differently, we wish to test whether the relative number of cross-bridges attached to the actin filaments at any one time is a function of the degree of filamentary overlap, the active state, *and* the interfilamentary velocity, or whether it is a function of the first two factors only. Evidently, we should be able to extract this information from observations of isometric twitch contractions executed at

$$\ell_o = \bar{\ell} - \bar{\alpha}\lambda_{so} \qquad (6.11)$$

where $k(\xi) \approx 1$. In this contractive mode the active state $q(\xi,\gamma)$ rises very rapidly, while the initially easily extendable tendon and cross-bridge elasticities permit a high initial shortening velocity $\dot{\xi}$ of the contractile element, implying a small initial force output.

If however, the assumption were made that the relative number $\Omega_{ik}$ of cross-bridges attached to the actin complex, were only a function of the degree of filamentary overlap and the active state, and *not* of the interfilamentary velocity (i.e. if $h(\dot{\xi}_{ik}) \equiv 1$ in (4.4)), the stiffness of the cross-bridge elasticities should increase rapidly with the rise of $q(\cdot)$, in accordance with (4.4). This, in turn, would imply a rapid reduction of the initial extendability of the combined cross-bridge and tendon elasticities, entailing a lower initial internal shortening velocity, and hence a phase of fast initial force rise. That simulation response which agrees with comparable experimental observations, will be considered as proving the respective hypothesis correct.

For the definition of variable names used below, see the listing of the program MYOSIM in Appendix A4.

## MODEL SPECIFICATIONS

*Simulation run control:*

MODEL = 1, IGRAPH = 1
TSTART = 0.0, TFIN = var., OUTINT = 0.001,
TAB = 0.0, HSTEP = 0.00005,
T1 = 0.0, T2 = 0.0, T3 = 0.0, T4 = 0.0, SLOPE1 = 0.0, SLOPE 2 = 0.0

For the simulation responses shown in the left diagrams of Fig. 6.1 the variable ICBEL in the main program had a value of 1. If not indicated otherwise, ICBEL always has a value of zero. The symbol 'var.' indicates that the value of the respective constant was changed for different simulation runs. The values actually used are given in the legends of the respective illustrations.

*Model parameters:*

N = 1, EL = var., ALFAB(1) = 0.08, ALSO(1) = 0.028, ELB(1) =
= 0.05,
IFF(1) = 1 (fast fibre)
      = 0 (slow fibre),
YO(1) = 0.0, YO(2) = 0.0, YO(3) = 0.0, YO(4) = 0.0, YO(5) = 1.0

*Neural control mode:*

FRE(1) = 1.0   (Single 1 ms nerve volley at t = 0)

*Contractive mode:*

Isometric at various fibre lengths.

*Graphic display mode:*

IFIB = 1, LINE = 1, NPLOTS = 2,
INDEX (1) = 4, INDEX(2) = 6

Some remarks concerning the selection of the values of the model parameters may be in order. The choice of the value of $\bar{\ell}(=\text{ELB})$ is arbitrary (0.05 m in the present case). However, the permissible values of

$\ell(=EL)$ and $\lambda_{so}(=ALSO)$ depend on $\bar{\ell}$. For any selected contractive mode, the fibre length $\ell$ must always lie between the physiological limits

$$\ell_{min} = 0.6\bar{\ell} + (0.4-0.5\bar{\alpha})\lambda_{so} \qquad (6.12)$$

and

$$\ell_{max} = 1.7\bar{\ell} - (0.7+1.7\bar{\alpha})\lambda_{so}, \qquad (6.13)$$

which ensures that the domain of definition of the filamentary-overlap function for single fibres is not exceeded. This restriction does not apply to the global model (Model 3).

The value of $\lambda_{so}$ determines the amount of tendinous material present in the fibre. Close (1965, p.551) gives values of 0.35 to 0.40 for the ratio between muscle fibre length and muscle rest length for a variety of muscles. This ratio is approximately equal to $\bar{\lambda}/\ell_o = 1-\lambda_{so}/\ell_o$. Hence on average,

$$\lambda_{so} \approx 0.6\ell_o = 0.6\bar{\ell}/(1+0.6\bar{\alpha}), \qquad (6.14)$$

where (6.11) has been used.

The value of $\bar{\alpha}$ can be estimated from data on the tensile properties of tendinous tissue supplied by Yamada (1970, Fig. 78, p. 100 and Table 73, p. 101). For man and most animals, and for the tendons of various muscles, Yamada found an ultimate percentage elongation of about 9.5–10. Owing to the nonlinearity of the stress-strain relation of tendons, the elongation corresponding to maximum isometric stress is approximately 7.5–8%. Hence an appropriate choice for the value of $\bar{\alpha}(=ALFAB)$ is 0.08.

With this value of $\bar{\alpha}$, and the value chosen for $\bar{\ell}$, the approximate value of $\lambda_{so}$ can be calculated from (6.14). For the present case this value is $\lambda_{so} \approx 0.028$, which was used in the simulation.

The initial state $YO(i)=0$, $i=1,\ldots,4$, $YO(5)=1$, corresponds to the resting state of the fibre, while $\ell=0.0479$ is the resting length $\ell_o$ of the fibre, as given by (6.11).

The results of the simulation runs are depicted in Fig. 6.1.

In the left-hand diagrams of Fig. 6.1 it is assumed that the number of myosin-actin links is *independent* of the interfilamentary sliding velocity and is influenced only by the active state and the degree of filamentary overlap. Since the mode of contraction is isometric at $\ell=\ell_o$, we have that $k(\xi)\approx 1$. Thus the hypothesis is reduced to the statement that crossbridge attachement is a function of the degree of activation only. As expected, the rapidly rising active state produces a rapid increase in the

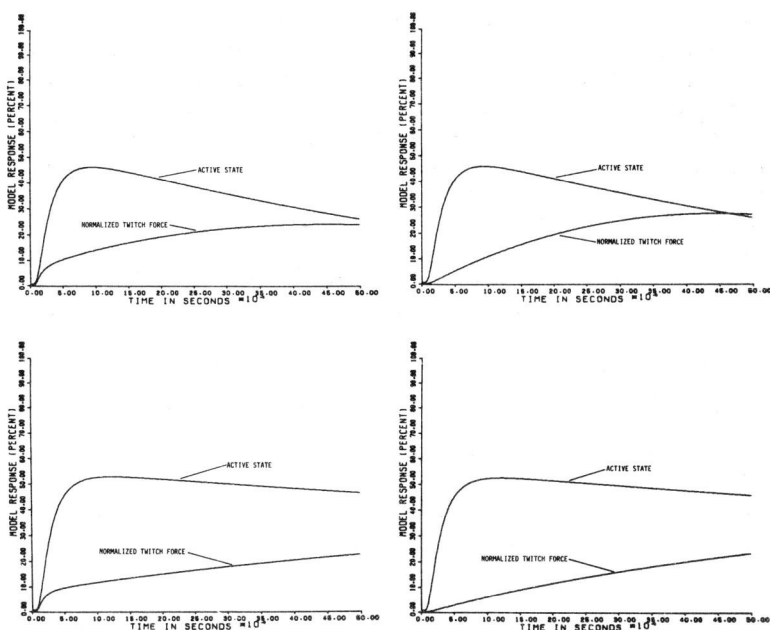

**FIG. 6.1:** Initial phases of twitch responses. Top diagrams: fast twitch fibre; bottom diagrams: slow twitch fibre. Left-hand diagrams: cross-bridge attachment assumed independent of sliding velocity; right-hand diagrams: cross-bridge attachment assumed dependent on interfilamentary sliding velocity. Simulation time TFIN = 50 ms and fibre length EL=$\ell_o$=0.0479 m. Output interval OUTINT=0.0002 s

number of cross-links, and hence a rapid increase in the stiffness of the cross-bridge elasticities, in accordance with (4.4). This effect manifests itself in a phase of rapid initial force rise which is clearly visible in the left-hand diagrams of Fig. 6.1, for both fast and slow twitch fibres.

On the other hand, the hypothesis of a dependence of the number of cross-links on the active state, the degree of filamentary overlap *and* the interfilamentary velocity generates the results shown in the right-hand diagrams of Fig. 6.1. As the active state rises, so does the interfilamentary velocity of the contractile structures that shorten against the series-elastic structures. Since cross-bridge attachment is postulated to depend on this velocity (via the function $h(\dot{\xi}_{ik})$ in (4.4)) the stiffness of the cross-bridge elasticities rises only slowly, producing the flat and smooth initial phase of initial force rise observable in the right-hand diagrams of Fig. 6.1.

All available experimental recordings of isometric twitch contractions of mammalian skeletal muscle at body temperature (Norris, 1961a, 1961b; Close, 1965; etc.) clearly show the initial force rise depicted in the graphs on the right-hand side of Fig. 6.1, while the pattern shown in the left-hand graphs has never been observed. One is thus led to conclude that the number of myosin-actin links formed is strongly dependent on the interfilamentary velocity, at least at negative velocities (shortening), and that the velocity-dependence of the force output of a muscle is predominantly due to this phenomenon, and not to an inherent property of the cross-bridge dynamics. At the same time it is clear that the substitution into Equation (4.4) of the function $h(\dot{\xi}_{ik})$ is appropriate.

In Fig. 6.2 are shown the full twitch responses at length $\ell_o$ of an average fast and an average slow-twitch fibre. The terms 'average fast' and 'average slow' are to be understood to refer to fibres that have twitch contraction times of 50 ms and 120 ms respectively.

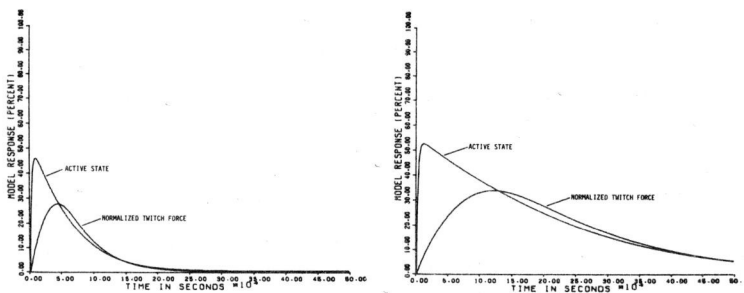

**FIG. 6.2:** Twitch responses of average fast (left-hand diagram) and average slow (right-hand diagram) fibres at $EL = \ell_o = 0.0479$ m. Simulation time TFIN = 500 ms

Fig. 6.2 shows that all the well-known characteristics of the isometric twitch are correctly predicted by the model: the *twitch contraction times* are about 47 ms and 120 ms respectively for the average fast and slow fibre, while the corresponding values for the *half-relaxation times* are 51 ms and 180 ms. These values agree well with experimental data reported in the literature (e.g. Eberstein and Goodgold, 1968).

Good agreement between model predictions and experimental results also holds for the *twitch-tetanus ratio:* about 0.28 for the average fast and 0.34 for the average slow fibre. Close (1972) reports a twitch-tetanus ratio of 0.2–0.3 for fast, and somewhat higher values for slow mammalian muscle fibres at 37°C. Moreover, the time course of the *active state* of the model muscle fibre agrees well with that determined experimentally on real muscle fibres (Bahler et al., 1967; Siegman and Gordon, 1972; etc). In comparing the active state of the model with experimental results it must, however, be kept in mind that the model active state is defined to be the relative amount of calcium bound to troponin. This is the definition proposed by Ebashi and Endo (1968, p. 139) which is conceptually different from that of Hill (1949).

Finally, the right-hand diagrams of Fig. 6.1 reveal that the *latency* of about 1.2 ms (Norris, 1961a) in the onset of the contractive force after stimulation is correctly predicted by the model for both average fast and slow fibre.

A crucial test for the validity of any muscle model is its capability of correctly predicting the peculiarities of the twitch responses at *various fibre lengths*. Fig. 6.3 shows for the fast muscle fibre the predicted active-state and twitch-force histories for the fibre lengths $\ell=0.042$ $(0.877\ell_o)$, $\ell=0.046$ $(0.960\ell_o)$, $\ell=0.054$ $(1.127\ell_o)$, and $\ell=0.058$ $(1.211\ell_o)$, while Fig. 6.4 displays superimposed twitch-force histories of the records shown in Fig. 6.3, with the passive force components removed.

In the two upper diagrams of Fig. 6.3 the fibre is at lengths less than its rest length $\ell_o = 0.0479$, and initially at rest. Upon stimulation, the contractile element shortens and extends the initially slack tendon. During this phase the externally observable force output is obviously equal to zero, which is clearly illustrated by the delayed onset of twitch forces in Fig. 6.3. The same effect has been observed frequently in experiments on mammalian and frog muscle (see, for example, Fig. 3 of Ritchie and Wilkie, 1958).

In order to obtain the genuine twitch response at short fibre lengths, the fibre is restimulated at the moment when the contractile force has just reached the zero value. This is the case at $t=0.2$ s in the left upper diagram, and at $t=0.263$ s in the right upper diagram of Fig. 6.3. Since the contractile element is now at a shorter length the peak of the active state is lower and its decay more rapid than that of the first twitch. Even so, the resulting twitch force is larger.

At fibre lengths greater than the rest length $\ell_o$, the passive force component becomes visible, as illustrated in the two lower diagrams of Fig. 6.3. It can also be seen that the active state is greatly enhanced at these lengths, both as far as its peak and its decaying phase are concerned, a fact which has been confirmed experimentally (Bahler et al., 1967).

To enable the normalized twitch forces at various fibre lengths to be compared, these force histories have been superimposed in Fig. 6.4, with the passive force components removed.

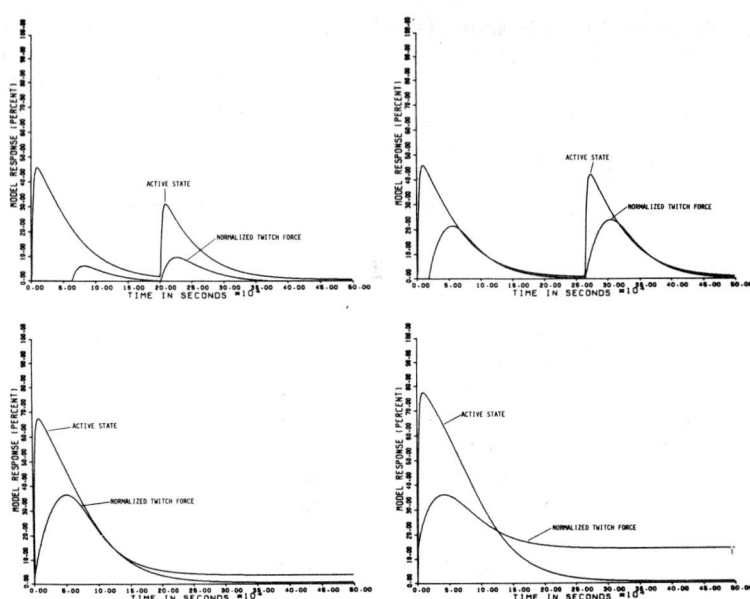

**FIG. 6.3:** Twitch responses at various fibre lengths. Active-state and twitch-force histories for the fibre lengths EL (from left to right and top to bottom): $0.877\ell_o$, $0.960\ell_o$, $1.127\ell_o$, and $1.211\ell_o$. Twitch forces have been normalized with respect to the maximum isometric force. Simulation time TFIN=0.5 s. Detailed explanations in the text

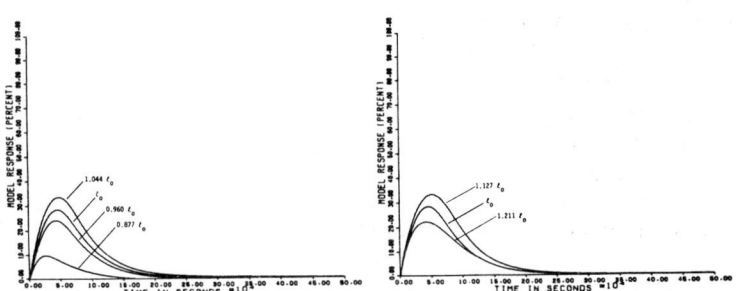

**FIG. 6.4:** Normalized twitch forces at various fibre lengths EL. Rest length $\ell_o = 0.0479$. Simulation time TFIN=0.5 s

It can be seen from Fig. 6.4 that all the important features of the twitch response at different fibre lengths are correctly predicted by the model: an increase of both twitch contraction time and maximum twitch force up to about $1.127\ell_o$ (8% above the optimum length $\bar{\ell}$) as well as a markedly delayed decrease of the twitch force at long lengths (see curve for $\ell = 1.211\ell_o$). A direct consequence of these observations is that the *length-tension curve for twitch contractions* has its peak at a length greater than the optimum length $\bar{\ell}$, and that the *half-relaxation time* of a twitch increases with the fibre length. In the present simulation the length-tension curve for twitch contractions has its peak at $\ell = 1.08\bar{\ell}$, while the half-relaxation times are 43.7 ms at $\ell=0.877\ell_o$, 49.2 ms at $\ell=0.960\ell_o$, 50.6 ms at $\ell=\ell_o$, 53.3 ms at $\ell=1.127\ell_o$, and 61.5 ms at $\ell=1.211\ell_o$. All of these observations are in agreement with experimental results obtained on frog and mammalian skeletal muscles (Rack and Westbury, 1969, Fig. 4; Bahler *et al.*,1967, Figs. 9–11; Buller *et al.*, 1960, Fig. 1; Stevens *et al.*, 1980, Fig. 3; etc.).

In comparing the simulation results with experimental observations it must, however, be kept in mind that the values of the constants used in the model have a pronounced effect on the simulation responses. This is particularly true for the constant $\tilde{\xi}$ appearing in (3.26), which determines the length-dependence of the active state. Its value used in the present simulation studies is 2.90, which appears appropriate for human muscle. A smaller value of $\tilde{\xi}$ ($\tilde{\xi} \geq 1.85$) enhances the effect of active-state potentiation at long lengths, while a larger value of $\tilde{\xi}$ diminishes it. Hence twitch responses of various muscle types can be simulated by appropriately adjusting the value of this parameter.

It should be noted that the values of the lengths in Fig. 6.4 are given as fractions of the rest length $\ell_o$ rather than the optimum length $\bar{\ell}$. The reason is that $\ell_o$ is the greatest length at which the passive force component is still zero. If required, the results are easily related to $\bar{\ell}$ by using expression (6.11).

This completes the simulation studies of isometric twitch responses of single fibres. In the next section we shall be concerned with model responses to repeated stimulation.

### 6.2.2 Simulation of multi-stimulative responses

If a muscle fibre is subjected to a series of successive neural input signals at varying frequencies, its response exhibits a highly nonlinear behaviour. We shall now demonstrate that all these nonlinearities are correctly predicted by the present model.

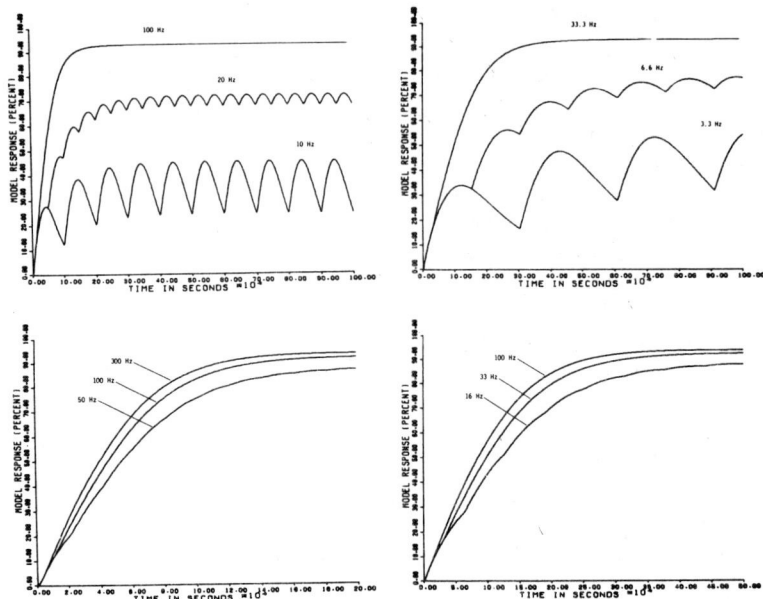

**FIG. 6.5:** Normalized force outputs of fast (left-hand diagrams) and slow (rigth-hand diagrams) muscle fibres under isometric conditions at EL=$\ell_o$ = 0.0479, and for the stimulation rates FRE(1) indicated in the diagrams. Simulation times: TFIN=1 s for upper diagrams, 0.2 s for left lower diagram, and 0.5 s for right lower diagram. Further parameter values are T1=T2=0.0, SLOPE1=0.0, SLOPE2=0.0, NPLOTS=1, no INDEX(2). Three successive simulations were initiated by three different data sets for each graph (see comments section in program MYOSIM)

## MODEL SPECIFICATIONS

*Simulation run control:*

MODEL = 1, IGRAPH = 1
TSTART = 0.0, TFIN = var., OUTINT = 0.001, TAB = 0.0,
HSTEP = 0.00005, T1 = var., T2 = var., T3 = 0.4 , T4 = 0.5,
SLOPE1 = var., SLOPE2 = var.

*Model parameters:*

N = 1, EL = var., ALFAB(1) = 0.08, ALSO(1) = 0.028, ELB(1) = 0.05,
IFF(1) =1 (fast fibre)
=0 (slow fibre),
YO(1) = 0.0, YO(2) = 0.0, YO(3) = 0.0, YO(4) $\doteq$ 0.0, YO(5) = 1.0

*Neural control mode:*

FRE(1) = var. (Stimulation by trains of 1 ms nerve volleys at varying frequencies.)

*Contractive mode:*

Isometric and non-isometric (lengthening and shortening at prescribed constant rates).

*Graphic display mode:*

IFIB = 1, LINE = 1, NPLOTS = var.,
INDEX(1) = 5, INDEX(2) = var.

In Fig. 6.5 are shown the predicted relative force outputs of fast and slow fibres, at a variety of stimulation rates and under isometric conditions.

The curves in Fig. 6.5 display the well-known behaviour of the force output of isometrically constrained muscle fibres stimulated at various rates (see, for example, Rack and Westbury, 1969, Fig. 6; Buller and Lewis, 1965, Fig. 2; etc.). It is especially noteworthy that the *maximum tension* reached is dependent *on the stimulation rate* in a highly *nonlinear* way. For the fast muscle fibre, an increase of the stimulation rate from 10 Hz to 20 Hz causes the relative output force to increase from 0.35 to 0.70, while an increase of the firing rate from 20 Hz to 100 Hz causes the relative tension to increase further by only 0.23 to 0.93. A similar situation prevails with the slow fibre. This phenomenon is, of course, well known from experimental studies on mammalian muscles.

In the lower part of Fig. 6.5 we investigate the change of the maximum rate of tension rise with the stimulation rate. When the maximum values of the time derivatives of the tension curves are plotted as functions of the stimulation frequencies it turns out that for both fast and slow muscle fibres the relationship is approximately logarithmic. Again, this prediction is in agreement with the experimental findings of Buller and Lewis (1965) on fast and slow cat muscle.

The results of an experiment designed to investigate the length-dependence of the tetanic force of the model muscle fibre are displayed in Fig. 6.6.

The final tensions reached in Fig. 6.6 reflect, of course, the length-tension relation of the single muscle fibre. It is, however, interesting that for the slow muscle fibre (rigth-hand diagram in Fig. 6.6) the delay times for the onset of force at shorter lengths (38 ms at $0.92\ell$ and 122 ms at $0.84\ell$) correspond very closely to those found by Rack and Westbury (1969, Fig. 10) for the slow cat soleus muscle at 37°C (35 ms and 120 ms respectively). This clearly illustrates that the delay phenomenon in slack muscle is correctly predicted by the present model.

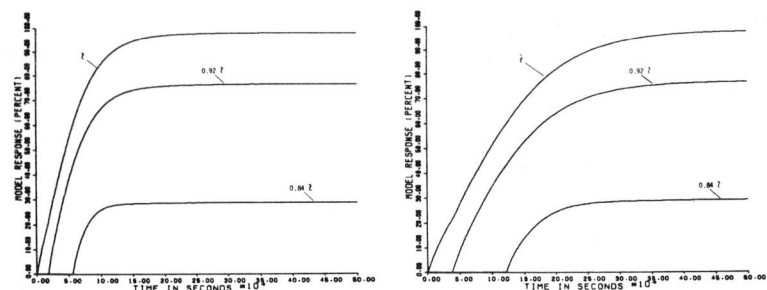

**FIG. 6.6:** Normalized force histories for fast (left-hand diagram) and slow (right-hand diagram) muscle fibres at the lengths EL indicated on the graphs ($\bar{\ell}=0.05$). Stimulation rates FRE(1) are 70 Hz for the fast and 25 Hz for the slow fibre. Further parameter values: TFIN=0.5, T1=T2=0.0, SLOPE1=0.0, SLOPE2=0.0, NPLOTS=1, no INDEX(2). Three successive simulations were initiated by three different data sets

In Fig. 6.7 are displayed tetanic responses of a fast muscle fibre, including the *posttetanic relaxation phases*. The onsets of the relaxation phases are indicated by arrows. A twitch is also shown, to facilitate comparison. In the right-hand diagram the posttetanic relaxation phases are superimposed in such a way that t=0 is the common switch-off time. These superimposed responses can be compared directly with the experimental recordings of Norris (1961 b, Fig. 3), made on rat peroneal muscle. All three phases of relaxation identified by Norris are

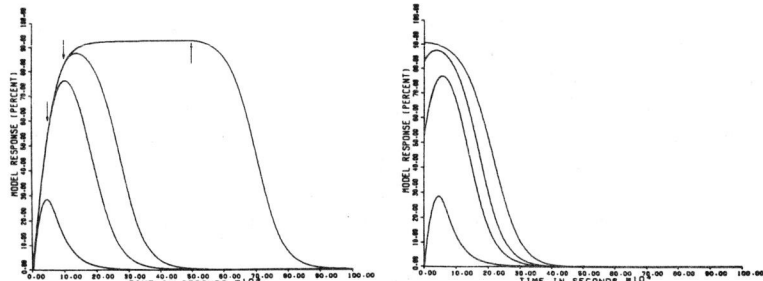

**FIG. 6.7:** Normalized isometric force outputs of the fast fibre, with the arrows in the left-hand diagram indicating the times of stimulation (0.05, 0.1, and 0.5 s respectively). The right-hand diagram shows the superimposed posttetanic relaxation phases of the responses shown in the left-hand diagram, and arranged so that t=0 represents the common switch-off time. Stimulation rate FRE(1)=100Hz, fibre length EL=$\ell_o$=0.0479. Other parameter values: TFIN=1, T1=T2=0.0, SLOPE1=0.0, SLOPE2=0.0, NPLOTS=1, no INDEX(2). Results of three successive simulation runs superimposed on twitch response

clearly visible in Fig. 6.7: a first phase of slow tension decay which is more prominent in full tetani than in twitches, a second phase of rapid tension decay, and a third phase of slower tension decay.

In terms of the present model these phenomena are explained as follows. In a supramaximal tetanus of long duration the interfilamentary calcium ion concentration $\gamma$ has almost reached its maximal level, which implies an almost maximal active state q. Upon termination of stimulation the Ca concentration decays radpidly, but owing to the plateau of the active state function $q(\xi,\gamma)$, the latter, and hence the isometric force, remains at a comparatively high level. If the duration of the tetanus was short, the Ca concentration did not reach a similarly high level and therefore did not penetrate far into the plateau region of the active state. The initial relaxation phase is then correspondingly shorter.

During the subsequent further decay of $\gamma$, the steep region of the active-state function $q(\xi,\gamma)$ is passed and a rapid force decline ensues which is only sligthly compensated for by the stretching action of the series-elastic element. This is the second relaxation phase of rapid force decline.

Finally, the (approximately exponential) decay of $\gamma$ slows down, and simultaneously the flat part of the active-state curve at low values of q has been reached. Both factors combine to produce the third relaxation phase of slow force decay. If the series-elastic structures present in the fibre are long compared with the contractile structures, then the length-tension relation also has a significant influence on the characteristics of the three relaxation phases.

We shall next discuss the most intriguing phenomenon of the *response of the muscle fibre to stretches*. In this contractive mode the force output of the fibre exhibits certain characteristics that are difficult to explain in terms of current theories of muscular contraction. Model studies involving this contractive mode may therefore provide valuable insight into these phenomena which could not be obtained otherwise.

If a tetanically contracting muscle fibre is stretched at a certain velocity, its contractile force rises rapidly at first to a level that depends on the stretching velocity and the fibre length, and then continues to increase, at a lower rate, during the remainder of the stretching period. Upon termination of the stretch the force decays, first rapidly and then more slowly, maintaining for a considerable time a level that is higher than the value of the isometric tension corresponding to the new length. After some time this extra tension then also disappears. The amount of extra tension *increases,* and its rate of decay decreases, with increasing muscle length beyond the optimum length $\bar{\ell}$. This is puzzling indeed, since at long lengths the degree of filamentary overlap decreases approximately linearly with increasing sarcomere length, which results in a corresponding *decrease* of the isometric tension and the initial stretching force (Hill, 1977; Edman *et al.*, 1978; Flitney and Hirst, 1978).

These effects have been observed most clearly in frog muscle fibres at low temperatures (Déléze, 1961; Sugi, 1972; Hill, 1977; Flitney and Hirst, 1978; Edman *et al.*, 1978; etc.) but are far less pronounced in mammalian muscles at 37°C (Stevens *et al.*, 1980; Joyce *et al.*, 1969).

According to the author's hypothesis (Hatze, 1973), stretches induce energy conversion processes in the cross-bridges such that part of the work done on the muscle fibre is temporarily stored in the myostructures, and released as heat upon termination of the stretch. At the onset of the stretch an immediate rise of the output force is predicted as a result of the reversal of the direction of cross-bridge movement, and the initiation of the energy conversion process (for details see Hatze, 1973). This process resembles the effect of a sudden displacement of a current-carrying coil in a magnetic field: the force resisting the displacement immediately rises, and the energy stored in the electric circuit of the coil increases. It is postulated that intermolecular forces between the two globular heads of the HMM – S1 subunit of the myosin molecule permit only minute excursions of the cross-bridge heads in the direction of lengthening. This would result in a rapid breaking of the actin-myosin link and a vigorous search for reattachment of the highly charged cross-bridge to the next available site. Once attached, the cross-bridge would again be moved in the direction of lengthening, which would further increase its electrical charge, and hence the energy stored.

The probability of reattachment is likely to be increased by the reduction, at increased fibre lengths, of the distance between the myosin and actin filaments. As is well known, this reduction is the result of the constant-volume relation (Dragomir, 1970). This facilitation has a two-fold effect: first, it reduces the 'inactive period' of search for an attachment site (Hatze, 1973, p. 228) resulting in a prolonged force-producing cross-linkage, and secondly it produces a proportionately larger increase of the energy stored in the cross-bridges owing to the longer period of attachment during the stretch. This twin-effect is highly nonlinear and therefore overcompensates for the reduction in the overall force output caused by the approximately linearly decreasing filamentary overlap. Indeed, a fairly accurate representation of this combined effect is given by the function

$$g_1(\xi) = 1 - k^2(\xi), \quad \text{for } \xi \geq 1,$$
$$= 0 \quad \text{otherwise,} \qquad (6.15)$$

where $k(\xi)$ is defined by (3.33).

At fibre lengths less than $\tilde{\ell}$, the separation between the actin and myosin filaments increases rapidly in accordance with the relationship $d = c'\xi^{-\frac{1}{2}}$, where $c'$ is a constant and $\xi$ is proportional to the sarcomere length. Hence the probability of attachment can be expected to decrease considerably, which reduces the *relative facilitation* $g_1(\dot{\xi})$ to negligible values, as is expressed by (6.15).

Under the present hypothesis the charge on a converter element (a cross-bridge) increases nonlinearly with the stretching velocity (Hatze, 1973, p. 237). This dependence can be expressed by the relation

$$g_2(\dot{\xi}) = h(\dot{\xi}) - 1/(1-q_o), \quad \text{for } \dot{\xi} \geq 0,$$
$$= 0 \quad \text{otherwise,} \qquad (6.16)$$

where $h(\dot{\xi})$ is given by (4.15). For $\dot{\xi}=0$, the function (6.16) vanishes.

It is thus clear that under the present assumptions the contractive state of the converter elements will be enhanced, in a time-dependent and highly nonlinear fashion, by an increase in both the stretching velocity and the fibre length, if the latter is greater than $\tilde{\ell}$. This enhancement constitutes a *potentiation, of a new type,* of the contractile structures, which is only partly due to a potentiation of the active state q (which we had defined to be the relative amount of Ca bound to troponin). Indeed, when the whole population of cross-bridges is considered, the overall potentation due to the stretch effect does, of course, also depend on the relative number of cross-bridges active, i.e. on the present active state q. The stretch potentiation nevertheless has the same effect as an enhancement of the active state itself. We shall, therefore, call it $\triangle q$. Its dynamic behaviour is described by a nonlinear second-order differential system (Hatze, 1973, p. 245) of which the following first-order differential equation represents an approximation:

$$\triangle \dot{q} = d_1\{d_2 q[1-k^2(\xi)][h(\dot{\xi})-1/(1-q_o)]-\triangle q\}, \quad \triangle q(t_s) = \triangle q_o. \quad (6.17)$$

In (6.17), $d_1$ and $d_2$ denote constants, while $t_s$ designates the time of stretch initiation. All other symbols have the meanings defined previously. The resting potentiation $\triangle q_o$ has a value of zero.

From (6.17) it can be inferred that the stretch potentiation and its rate of increase or decrease at any moment of time are determined by the present active state q, the degree of facilitation as defined by (6.15), and the stretching velocity $\dot{\xi}$.

We now define the *relative stretch potentiation* $\delta q$ by

$$\delta q \triangleq \triangle q/q, \qquad (6.18)$$

so that with $\delta \dot{q} \triangleq \triangle \dot{q}/q$, (6.17) becomes

$$\delta \dot{q} = d_1 \{d_2[1-k^2(\xi)][h(\dot{\xi})-1/(1-q_o)]-\delta q\}, \quad \delta q(ts) = \delta q_o. \quad (6.19)$$

The *total active state* q* of the muscle fibre is then given by

$$q^*(t) = q(t)(1+\delta q(t)), \qquad (6.20)$$

where q(t) is defined by (3.29).

It is important to note that by virtue of (6.15) and (6.16), Equation (6.19) becomes operative only when $\dot{\xi} > 0$ and $\xi > 1$. Since after a stretch the value of δq always decays to zero, the value of δq(t) in (6.20) is non-zero for only a limited period of time after a stretch has been induced at a fibre length corresponding to a value of $\xi > 1$.

There is a twofold reason why the differential equation (6.19) has not been included as part of the excitation dynamics (3.21–3.23), or (3.27). First, mammalian muscle *in situ* is seldom extended far beyond its optimum length $\tilde{\ell}$ in the body, and hence the value of δq(t) in (6.20) remains at a negligible level, as will be demonstrated below. Secondly, stretch potentiation due to an increase in δq is far less pronounced in mammalian muscle at 37°C than it is in frog muscle near 0°C. This is clearly demonstrated by the results of Joyce et al. (1969, Fig. 6) on cat soleus muscle at body temperature. It must, however, be emphasized that this does *not* mean that there is no initial force rise during the stretch in mammalian muscle. The initial force rise and its persistence during the stretching period are independent of the relative stretch potentiation δq, and are correctly predicted by the present model, as will be shown below.

Since the models developed in the present monograph are intended to represent mammalian muscle at 37°C, the effect of δq(t) on the force output of the model muscle can be neglected, and the differential equation (6.19) may be omitted from the excitation dynamics. In this case it follows from (6.20) that $q^*(t) \approx q(t)$. If, however, simulation of stretches at muscle lengths considerably greater than $\tilde{\ell}$ are performed, the influence of δq(t) may become appreciable, making it necessary to include (6.19) in the excitation dynamics (3.21–3.23) and to substitute (6.20) for the active state q. The corresponding augmentation has, of course, also to be performed for the differential system (4.37), where the function ε is to be replaced by

$$\varepsilon^* = \varepsilon(1+\delta\varepsilon), \qquad (6.21)$$

where the new state variable δε is defined by

$$\delta\dot{\varepsilon} = f^*(\xi,\dot{\xi},\delta\varepsilon), \qquad (6.22)$$

the function f*(·) denoting the right-hand side of (6.19).

We shall now proceed with the simulation of stretch responses of the fast fibre. The values of the constants $d_1$ and $d_2$ appearing in (6.19) were chosen as $d_1 = 5.0$ and $d_2 = 6.8$. It must, however, be emphasized that these values only serve to demonstrate the effect of the stretch potentiation but do not necessarily correspond to actual values for fast mammalian muscle.

In Fig. 6.8 are shown 3 mm – stretches (6% of fibre length) at various muscle lengths and stretching rates (see legend of Fig. 6.8). It can be seen that all the peculiar characteristics of the stretch response are correctly predicted by the model: the initial rapid force rise, the subsequent slower rise, the rapid decline of the force immediately after the termination of the stretch, and the subsequent slow decay of a tension which is considerably larger than the corresponding tetanic value, and

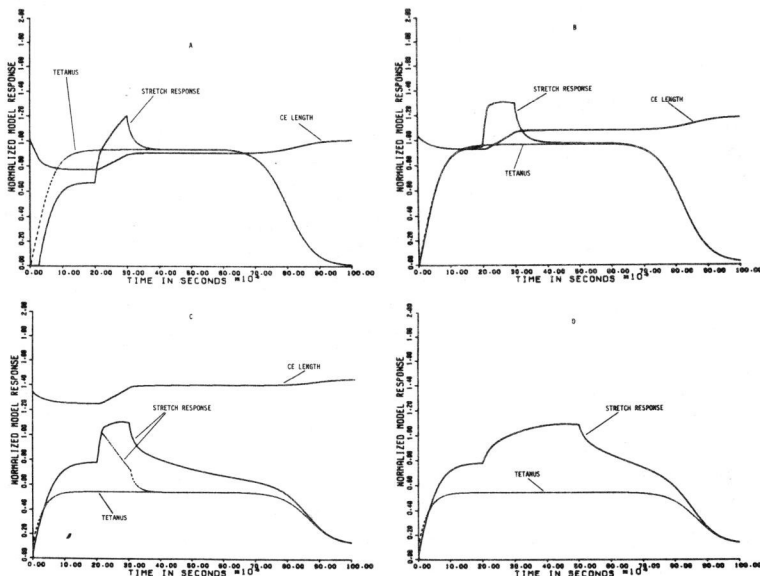

**FIG. 6.8:** Normalized force and length histories relating to stretches performed on a fast fibre. A: 3 mm – stretch at SLOPE1=0.03 m.s$^{-1}$ extending fibre from EL=0.045 (0.9$\tilde{\ell}$) to 0.048 (0.96$\tilde{\ell}$); also shown are tetanus at terminal length EL=0.048 and history of the length $\xi(t)$ of the CE. B: 3 mm – stretch at same rate as in A but starting at EL=0.0485 (0.97$\tilde{\ell}$) and terminating at EL=0.0515 (1.03$\tilde{\ell}$). C: 3 mm – stretch at same rate as in A from EL=0.0545 (1.09$\tilde{\ell}$) to 0.0575 (1.15$\tilde{\ell}$); dotted curve corresponds to a stretch without potentiation (i.e. with δq=0). D: 3 mm – stretch at rate SLOPE1=0.01 from EL=0.0545 to 0.0575. All tetani correspond to the terminal length of the stretch. Further parameter values: TFIN=1, OUTINT=0.002, HSTEP=0.0001, T1=0.2, T2=0.3 (0.5 in D), SLOPE2=0.0, FRE(1)=100, NPLOTS=2 (1 for D), INDEX(2)=3 (0 for D), LINE=0. In all simulations, stimulation is terminated at t=0.6 s. Total simulation time is 1 s

which gives rise to the phenomenon of force enhancement after stretch. What is most important, the model correctly predicts that this force enhancement should increase with increasing muscle length above $\bar{\ell}$. Note also that the predicted initial decay after termination of the stretch is slower after a slow stretch (Fig. 6.8D) than after a fast one (Fig. 6.8C), a fact which is also confirmed by experimental results on frog muscle (Déléze, 1961, Fig. 2; Edman *et al.*, 1978, Fig. 2).

Just how much the stretch potentiation enhances the force after a stretch at great length can be seen from Fig. 6.8C, where the upper dotted line represents the stretch response of the model without potentiation, i.e. with $\delta q \equiv 0$ in (6.20).

It is also of interest to compare the magnitude of the predicted initial force rise for different lengths of the contractile element. At $\xi = 0.78$ (Fig. 6.8A), which corresponds to a sarcomere length of 1.64 $\mu$m in frog muscle, the magnitude of the initial force is 63% of that at $\xi = 0.933$ (Fig. 6.8B). At $\xi = 1.243$ (Fig. 6.8C), which is equivalent to a sarcomere length of 2.61 $\mu$m, the initial force value is 87% of that at $\xi = 0.933$, although the isometric force values at $\xi = 0.78$ and $\xi = 1.243$ are about the same and equal to 61% of the maximum isometric force. Again, these predictions are in complete agreement with the experimental findings of Edman *et al.* (1978, Fig. 4A), who report a shift of the peak of the initial stretch-tension curve towards larger values of the sarcomere length.

Finally, the predicted functions $\xi(t)$ of the CE length as displayed in Figs. 6.8A – C, mimic almost exactly the corresponding experimentally observed histories of the sarcomere length during development of the tetanus and the subsequent stretch as reported by Edman *et al.* (1978, Fig. 5).

One important point needs to be stressed though: the characteristic features of the initial and subsequent force rise during stretch, and the phenomena occurring after its termination, are only *partly* due to the stretch potentiation induced by $\delta q$. This is demonstrated most clearly in Fig. 6.9, where the dotted lines represent the model response with $\delta q \equiv 0$, for $\ell = 1.03\bar{\ell}$ (Fig. 6.9A) and $\ell = 1.06\bar{\ell}$ (Fig. 6.9B). As can be seen, the differences between the two model responses are comparatively small at these lengths, indicating that the characteristic features of the stretch are well predicted by the model that does not contain the differential equation (6.19).

Finally, we demonstrate that the model correctly predicts stretch responses to small stretching velocities (Fig. 6.10A), and responses to controlled release which follows a stretch (Fig. 6.10B). It is interesting to note that the predicted stretch response depicted in Fig. 6.10B is almost identical with that observed experimentally on cat soleus muscle

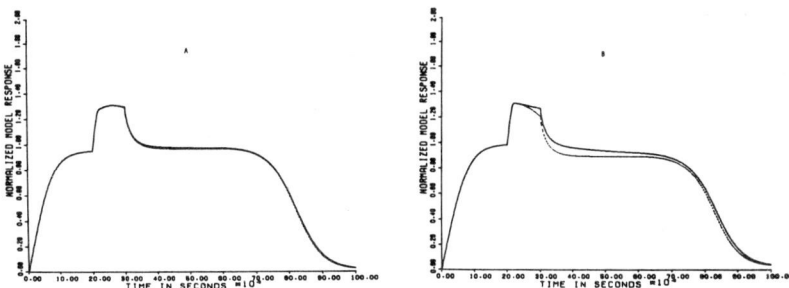

FIG. 6.9: Comparison of stretch responses with and without stretch potentiation. A: 3 mm – stretch at a rate of SLOPE1=0.03 from EL=0.0485 to 0.0515. B: same stretch, but from EL=0.050 to 0.053. In both diagrams: solid curves for stretch potentiation induced by δq, dotted curves for δq ≡ 0. Parameter values as for Fig. 6.8. Stimulation is terminated at t=0.6 s, and total simulation time is 1 s

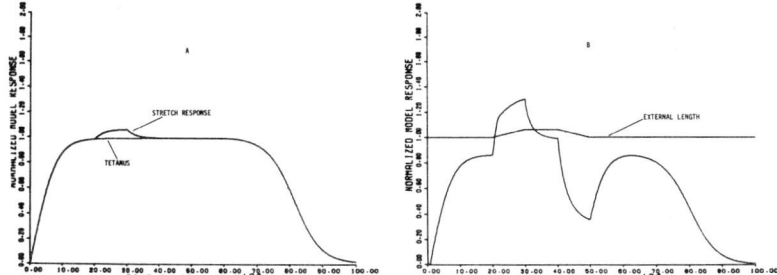

FIG. 6.10: Normalized force and length histories of responses to stretch and release. A: 0.3 mm – stretch at a low rate of SLOPE1=0.003 from EL=0.050 to 0.0503; tetanus (dotted line) at EL=0.0503. B: 3 mm – stretch with δq ≡ 0 in (6.20) at a rate of SLOPE1=0.03 from EL=0.047 (0.94$\bar{\ell}$) to 0.050 ($\bar{\ell}$), starting at T1=0.2 and terminating at T2=0.3; at T3=0.4 a release at a rate of SLOPE2=-0.03 ensues and terminates at T4=0.5, after which the isometric condition at EL=0.047 is maintained; the normalized external length change imposed on the fibre is also shown. In both diagrams stimulation ceases at t=0.6 s. Simulation time is 1 s. Other parameter values as for Fig. 6.8

(Joyce et al., 1969, Fig. 6) under similar conditions (tetanic stimulation, same starting length, relative stretching velocity, relative amount of stretch, and body temperature), while the predicted response to controlled shortening and subsequent redevelopment of tension (Fig. 6.10B) is precisely that observed on living muscle (Joyce et al., 1969, Fig. 1; Déléze, 1961, Fig. 3A). Note that the simulation depicted in Fig. 6.10B was performed with δq ≡ 0 in (6.20), i.e. without stretch potentiation.

117

In conclusion, we can state that the predictions of the present model, with $\delta q \equiv 0$ in (6.20), are sufficiently accurate for mammalian muscle at 37°C. Indeed, Joyce et al. (1969, p. 469) observed that in stretches 'the tension always rose as lengthening began, and always fell immedi-

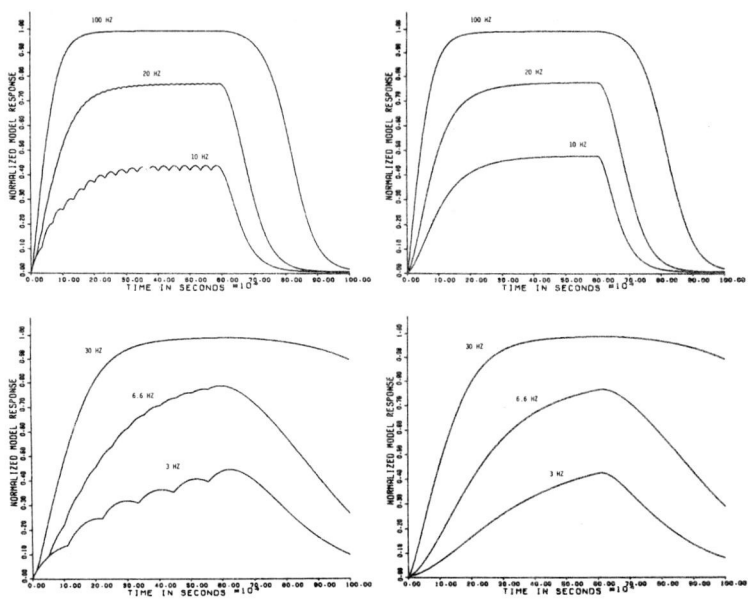

**FIG. 6.11:** Isometric total force responses of three simultaneously acting and asynchronously stimulated fast (upper left-hand diagram) and slow fibres (lower left-hand diagram), and corresponding responses of single 'equivalent' fibres of MODEL 2 (right-hand diagrams). The stimulation rates are indicated in the diagrams. The changes in the model specification parameters necessary to simulate MODEL 2 are obvious. Fibre length is $\bar{\ell}=0.05$, stimulation is terminated at $t=0.6$ s, and total simulation time is 1 s

ately the movement ceased', indicating that stretch potentiation in mammalian muscle at body temperature is much less pronounced than in frog muscle, at least at lengths not much exceeding $\bar{\ell}$. The above observations thus justify the omission of the differential equation (6.19) from the excitation dynamics of the fibre.

We have not yet verified that the 'lumped' excitation model (3.27) is, in fact, representative of an 'average' fibre of a muscle whose fibres are stimulated asynchronously. In Fig. 6.11 are compared the total tetanic force outputs of three simultaneously acting fast and slow fibres stimulated asynchronously at various frequencies, and the corresponding force outputs of MODEL 2 which uses (3.27) as excitation dynamics for a single 'equivalent' fibre. The asynchronous stimulation of the three fibres is achieved by exciting each of the N=3 fibres within a given stimulus interval τ once, but with a delay of $\triangle t$, where

$$\triangle_i t = (i-1)\tau/N, \quad i = 1,2,3. \tag{6.23}$$

This results in a total force output that is smoother than with synchronous stimulation, as has been shown experimentally by Rack and Westbury (1969, Fig. 5), and is also obvious from Fig. 6.11.

As can be seen, the responses of MODEL 2 (right-hand diagrams) provide reasonable approximations to those of MODEL 1 (left-hand diagrams), for both the fast and the slow fibre.

When the responses of MODEL 2 (right-hand diagrams of Fig. 6.11) are compared with those of MODEL 1 (left-hand diagrams of Fig. 6.11) it becomes apparent that the discrepancy between the responses of the two models is greater at low than at high stimulation rates. This is not surprising, since at low firing frequencies the force output of MODEL 1 consists of *unfused* twitch responses of the three fibres, and therefore contains distinct phases of contraction and relaxation, which clearly influence the course of the subsequent response. On the other hand, the force output of MODEL 2 does not contain such phases and hence must necessarily have a history different from that of MODEL 1. As the firing frequencies increase, the twitch tensions of MODEL 1 become fused, and the discrepancy between the responses of the two models becomes smaller.

In the living muscle, the fibres of many asynchronously firing motor units are intermingled. This means that even at low individual stimulation rates the total force output is smooth, resembling that displayed in the right-hand diagrams of Fig. 6.11.

In the next section we shall deal with the responses of the whole-muscle model.

## 6.3 COMPUTER EXPERIMENTS ON MODEL MUSCLES

So far we have investigated only responses of individual model fibres. In real muscles, however, fibres are always grouped into motor units of varying sizes. As discussed in Chapter 4, the histochemical, morphological and contractile profiles of these units are all different, which made it necessary to develop the global myocybernetic model (MODEL 3).

A validation of this model against corresponding experimental results is not possible, since to the best of the author's knowledge there are no records of experimental studies involving the responses of an intact muscle to artificially induced stimulations controlling separately each of its motor units. A somewhat crude comparison is nevertheless possible between this model (MODEL 3) and MODEL 4 consisting of two slow and one fast motor unit. Since the contraction dynamics of the global model is essentially the same as that of the single-fibre model, it is only necessary to validate the excitation dynamics (including the recruitment dynamics) of the former. This has been done in a simulation, the result of which is shown in Fig. 6.12.

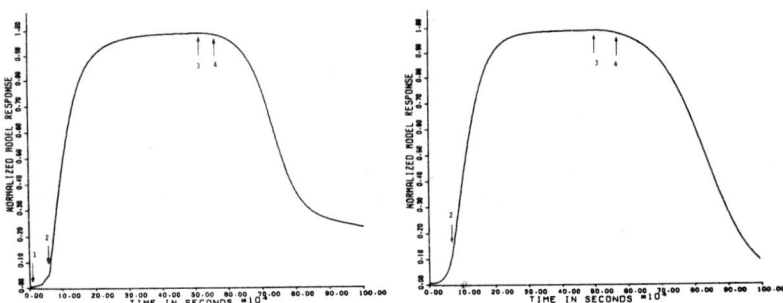

**FIG. 6.12:** Comparison of normalized isometric force responses of three motor units (left-hand diagram) and the model muscle (rigth-hand diagram). Arrows 1 and 2 indicate respectively beginning and end of motor unit recruitment, while arrows 3 and 4 mark respectively beginning and end of motor unit derecruitment (in the right-hand diagram arrow 1 coincides with t=0). Simulation time is 1 s and muscle length EL=0.05. The remaining parameter values are given in the text

The specifications for MODEL 3 were as follows:

*Simulation run control:*

MODEL = 3, IGRAPH = 1,
TSTART = 0.0, TFIN = 1.0, OUTINT = 0.002, TAB = 0.0, HSTEP = 0.0001, T1 = 0.0, T2 = 0.0, T3 = 0.0, T4 = 0.0, SLOPE1 = 0.0, SLOPE2 = 0.0

*Model parameters:*

N = 1, EL = 0.05, ALFAB(1) = 0.08, ALSO(1) = 0.028, ELB(1) = 0.05
YO(1) = 0.0, YO(2) = 0.0, YO(3) = 0.0, YO(4) = 0.0, YO(5) = 1.0

*Neural control mode:*

$V = 1.0, Z = 1.0$   for $0 \leqslant t \leqslant 0.07$
          = 0.0   for $0.07 < t \leqslant 0.50$
          = -1.0   for $0.50 < t \leqslant 0.57$
          = 0.0   for $0.57 < t$

*Contractive mode:*
isometric at $\bar{\ell}$
*Graphic display mode:*
IFIB = 1, LINE = 1, NPLOTS = 1, INDEX(1) = 6
The specifications for MODEL 4 have the format of MODEL 1 and are similar to those of MODEL 3, except for the *following parameters:* MODEL = 4, N = 3, IFF(1) = IFF(2) = 0 (two slow motor units), IFF(3) = 1 (one fast motor unit), FRE(1) = FRE(2) = 30.0, FRE(3) = 100.0.

*The recruitment times were:* MU(1) at t=0.01166, MU(2) at t=0.035, and MU(3) at t=0.05834, while derecruitment for MU(3) began at t=0.51166, for MU(2) at t=0.535, and for MU(1) at t=0.55834. This arrangement ensures an evenly distributed recruitment (derecruitment) pattern over the recruitment (derecruitment) interval of 0.07 s.

A comparison between the left and rigth-hand diagrams of Fig. 6.12 clearly reveals the differences between the responses of a distributed model consisting of only three motor units, and a lumped model consisting of a very large number of motor units. While the initial force rise up to about 0.1 s is about the same for both models owing to the exponential increments in motor unit size during recruitment and the comparatively small active state, the subsequent phase of the force rise is different for the two models. At first, the response of MODEL 4 (left-hand diagram of Fig. 6.12) rises faster, but then more slowly than that of MODEL 3. A similar pattern can be observed in the relaxation phases that follow the termination of stimulation as indicated by arrow 4. The reason for these discrepancies is, of course, the difference between the two models in the distribution of motor unit properties. While MODEL 4 comprises two average slow units and one average fast unit, the properties of the motor units of MODEL 3 range continuously over the whole spectrum, from the slowest to the fastest unit. This feature causes the response shown in the right-hand diagram of Fig. 6.12, a response closely resembling that found in living whole skeletal muscle (Hatze, 1978a, Fig. 4).

As mentioned previously, the contraction dynamics of the whole-muscle model is the same as that of the single fibre, and hence need not again be validated.

This completes the validation of the four myocybernetic models of skeletal muscle described in this chapter. In the next, we shall discuss some of the many possible applications of these models, with particular emphasis on the global model.

# CHAPTER 7

# Applications

Models are created for a purpose, and the myocybernetic models of skeletal muscle developed in this book are intended to be applied to the simulation of the dynamic behaviour and the study of the optimality principles governing the functions of neuro-musculoskeletal control systems. The discipline concerned with the investigation of these principles has been termed 'myocybernetics'. We shall say more about this new field in Section 7.2.

In the simplest case, the control system consists of a single muscle fibre contracting isometrically in response to a single nerve volley. In the most complicated case the system comprises the entire skeletal, muscular and neural assemblage that constitutes the human body.

This chapter will be devoted to a description of a model of the total human neuro-musculoskeletal system, to the application of the model to a real-life problem, and to the exposition of the foundations of myocybernetics.

## 7.1 HOMINOID DYNAMICS

The validations and simulation studies carried out in the previous chapter involved isolated model muscles and model fibres that were subjected to prescribed contractive modes and control inputs. Normally, however, a skeletal muscle in the intact body contracts against inertial loads, gravitational forces, and forces produced by other muscles and passive viscoelastic structures. The dynamic behaviour of the muscle therefore influences the state of the whole system and, in turn, is influenced by the state of the system of which it is an integral part. It is thus no longer possible to consider the dynamics of a muscle situated in the body in isolation, and a systems approach becomes necessary.

## 7.1.1 Definitions

We shall now define certain concepts that will be needed in the sequel.

A *neuro-musculoskeletal control system* is defined to be *a skeletal assemblage whose elements are acted upon by myo-(muscular) actuators which, in turn, are controlled by neural input signals* (Hatze, 1980b). The present treatment will be concerned exclusively with the important case of the human neuro-musculoskeletal control system. In this case, the assemblage of limb segments constitutes the *skeletal, or executor, subsystem* while the assemblage of striated muscles driving the skeletal subsystem is to be identified with the *subsystem* of *myoactuators*. The inputs directly controlling the actuator outputs are the neural signals emanating from the motoneurons in the spinal cord and the brainstem, i.e. from the *controller (neural) subsystem*.

A *hominoid* is an anthropomorphic mathematico-geometrical model of the segmented human body (Hatze, 1980c). It represents an intermediate stage in the transition from the real biosystem (the human body) to the abstract mathematical model given by the set of differential equations that describe the dynamic behaviour of the neuro-musculoskeletal control system (abbreviated NMSCS). In fact, the first step in deciding on the complexity of a human-body model is the selection of an appropriate hominoid consistent with the intended purpose of the model. The morphology of the hominoid defines the number, f, of mechanical degrees of freedom of the (unconstrained) model, as well as the shapes and inertial properties of the individual segment models.

Hominoids of widely varying complexity have appeared in the literature, beginning with the impressive work of O. Fischer in 1906. The simplest models currently in use, which could conceivably still be classified as anthropomorphic, consist of a three-link assemblage of uniform rigid bodies (trunk with two stiff legs, without feet or knee joints), and in some cases even of a single inverted pendulum with only one degree of freedom (Hemami *et al.*, 1973, 1977). These models are used to theoretically investigate biped stability of robots and manipulator locomotion systems, and employ the actuator torques as controllers of the model dynamics. For this reason and because they are extremely simple, these models are not of great value in studies involving more complex motions of the NMSCS.

Vukobratovič and Juricic (1969) have proposed a four-segment model for investigations into bipedal gait, while Hemami and Farnsworth (1977) used a five-segment model (trunk, left and right thighs and legs which include the feet) for the analysis of planar postural and gait stability.

Hominoids comprising from 11 segments (Morecki *et al.*, 1975), to 17 segments (Hatze, 1977b, 1980d) have been proposed for simulating and

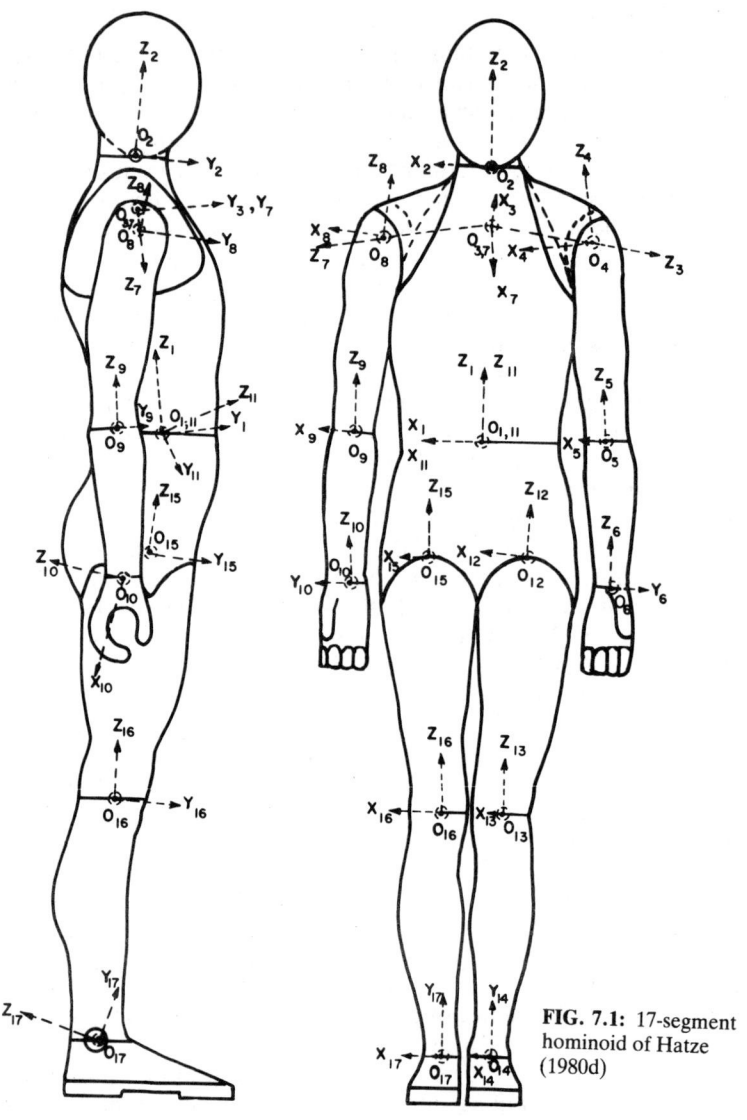

**FIG. 7.1:** 17-segment hominoid of Hatze (1980d)

analyzing more complex motions of the human NMSCS (for a detailed discussion see Hatze, 1980b). Surprisingly, however, the shoulder segments, which clearly constitute dynamically separate entities, are almost always considered part of the upper trunk except in (Hatze, 1980d), where they are modelled as separate segments.

The number, f, of mechanical degrees of freedom of the hominoids discussed varies from one (inverted pendulum in planar motion) to 44 (for three-dimensional motions of the author's 17-segment hominoid depicted in Fig. 7.1).

The evolution in configuration space (which is a subspace of the state space) of the solution of the differential system which describes the dynamics of the NMSCS, can be made visible by an equivalent representation of the actual motions of the skeletal and muscular subsystems of the hominoid. Such a graphical representation will be given later in this chapter (see Fig. 7.3). We are thus justified in using the term 'hominoid dynamics' to describe the dynamical behaviour of the NMSCS-model.

### 7.1.2 The model of the executor subsystem

As defined in the previous section, the (human) executor subsystem of the NMSCS consists of the assemblage of limb segments of the human body. The individual elements of this assemblage are connected by joints of various types in a tree (open-chain) structure.

The dynamics of this assemblage is described by a set of nonlinear, ordinary second-order differential equations which are obtained by applying Newtonian mechanics (or d'Alembert's principle), or the Lagrangian formalism (Hatze, 1977b). Using the latter and assuming that the assemblage consists of n individual elements (segments) admitting a total of f mechanical degrees of freedom described by the f-vector q, the equations of motion are given by

$$\frac{d}{dt}\left[\frac{\partial L}{\partial \dot{q}_i}\right] - \frac{\partial L}{\partial q_i} = Q_i, \quad i=1,\ldots,f. \tag{7.1}$$

where the Lagrangian L is defined (Hatze, 1977b) by

$$L = \sum_{i=1}^{n} M_i((\tfrac{1}{2})\dot{\rho}_i \cdot \dot{\rho}_i - gk \cdot \rho_i) + (\tfrac{1}{2})\omega_i \cdot \bar{I}_i \cdot \omega_i, \tag{7.2}$$

and $Q_i$ are non-conservative, applied generalized forces (actual forces or moments, depending on the nature of $q_i$) acting on the members of the assemblage. We shall elaborate on the $Q_i$ later. In (7.2), the symbols $M_i, \rho_i, g, k, \omega_i$ and $\bar{I}_i$ denote respectively the mass, the inertial position vector of the mass centroid, the gravitational constant, the unit vector in the direction of the z-axis, the inertial angular velocity vector with components relative to segment-fixed axes, and the inertia tensor of the i-th segment. The inertia tensor is assumed to be given in diadic form.

By carrying out the time differentiation in (7.1) we obtain

$$\frac{\partial^2 L}{\partial \dot{q}_i \partial \dot{q}_j} \ddot{q}_j = \frac{\partial L}{\partial q_i} - \frac{\partial^2 L}{\partial \dot{q}_i \partial q_j} \dot{q}_j + Q_i, \qquad i,j=1,\ldots,f, \qquad (7.3)$$

where summation over repeated indices is implied. Note that

$$\frac{\partial^2 L}{\partial \dot{q}_i \partial \dot{q}_j}$$

is a symmetric and positive definite $f \times f$ matrix.

We now define the 2f-dimensional state vector x by

$$x \triangleq (q,\dot{q})^T, \qquad (7.4)$$

so that the differential system becomes a first-order one given by

$$\dot{x}_j = x_{f+j}, \qquad x_j(0) = x_j^o, \qquad (7.5)$$

$$\frac{\partial^2 L(x)}{\partial x_{f+i} \partial x_{f+j}} \dot{x}_{f+j} = \frac{\partial L(x)}{\partial x_i} - \frac{\partial^2 L(x)}{\partial x_{f+i} \partial x_j} x_{f+j} + Q_i, \qquad x_{f+j}(0) = x_{f+j}^o$$

for $i,j=1,\ldots,f$, or more concisely by

$$\dot{x} = A^{-1}(x)[B(x)+Q], \qquad x(0) = x_o, \qquad (7.6)$$

where the symbols have the following meaning: $A^{-1}(x)$ denotes the inverse of the $2f \times 2f$ matrix

$$A(x) \triangleq \begin{bmatrix} I_f & 0_f \\ \hline 0_f & \dfrac{\partial^2 L(x)}{\partial x_{f+i} \partial x_{f+j}} \end{bmatrix} \qquad (7.7)$$

with $I_f$ and $0_f$ designating respectively the $f \times f$ identity and zero matrix. The 2f-vectors $B(x)$ and $Q$ are respectively defined by

$$B(x) \triangleq \begin{bmatrix} x_{f+1} \\ \vdots \\ x_{2f} \\ \hline \dfrac{\partial L(x)}{\partial x_1} - \dfrac{\partial^2 L(x)}{\partial x_{f+1} \partial x_j} x_{f+j} \\ \vdots \\ \dfrac{\partial L(x)}{\partial x_f} - \dfrac{\partial^2 L(x)}{\partial x_{2f} \partial x_j} x_{f+j} \end{bmatrix}, Q \triangleq \begin{bmatrix} 0 \\ \vdots \\ 0 \\ \hline Q_1 \\ \vdots \\ Q_f \end{bmatrix} \qquad (7.8)$$

The last quantity appearing in (7.6), but not yet discussed, is the vector Q of *applied generalized forces* acting on the system. This vector can be decomposed into the sum

$$Q = Q^M + Q^L + Q^E + Q^C, \qquad (7.9)$$

where $Q^M$, $Q^L$, $Q^E$, and $Q^C$ denote the column vectors of muscle (actuator) moments, joint range constraint moments, external generalized forces (such as wind resistance), and generalized external constraint forces, respectively. As is evident from (7.8), the first f components of these vectors are zero.

The vector $Q^M$ of myoactuator moments will be discussed later, while the vector $Q^E = Q^E(x,t)$ needs no explanation (gravitational moments and forces are, of course, already included in $B(x)$). The vector $Q^L$ accounts for passive (elastic and viscous) moments created by the *passive* structures (ligaments, connective tissue, resting muscles) spanning a joint. The component functions $Q_i^L(x)$, $i = f+1,\ldots,2f$, are highly nonlinear, as demonstrated experimentally in (Hayes and Hatze, 1977) for flexing movements of the human elbow joint, and automatically restrict the range of joint movements to the anatomical range.

Finally, the vector $Q^C$ of external constraint forces needs special mention. If the motion of parts of the hominoid is constrained by environmental conditions (such as ground contact of the feet in locomotion) then holonomic, and possibly rheonomic, state inequality constraints of the kind

$$f_j(x,t) \leq 0, \qquad j=1,\ldots,r, \qquad (7.10)$$

become operational. At the boundary of the region (7.10), i.e. upon activation of the constraint (for instance at the instant just after heel strike in walking) we have that

$$f_j(x,t) = 0, \qquad j=1,\ldots,r, \qquad (7.11)$$

which means that some of the coordinates describing the configuration of the hominoid become superfluous and could be eliminated. In other words, the dimension of the state space would be fluctuating on the different subarcs of the state trajectory, clearly an undesirable situation. This can be prevented by the appropriate introduction of constraint forces $Q^C(t)$ that are such that (7.11) is satisfied for the time of constraint activation. For $f_j(x,t) < 0$, $j=1,\ldots,r$, we have that $Q^C(t) = 0$. The problem of the computation and numerical stabilization of appropriate vector functions $Q^C(t)$ is discussed in detail in (Hatze, 1980c). It should be noted that the computation of the constraint force vector $Q^C(t)$ automatically yields all ground reaction force components, which eliminates the problem of indeterminacy of these forces, a problem that has bedevilled researchers in this field.

### 7.1.3 The model of the myoactuator subsystem

It is obvious that the vector $Q^M$ of myoactuator moments appearing in (7.9) provides the connection between the myoactuator subsystem and the executor (skeletal) subsystem. This connection will now be made clear.

Skeletal muscles in the body span one or more joints as is illustrated in Fig. 7.2 for a one-joint muscle.

Assume that the joint under consideration has three degrees of freedom and a non-stationary centre of rotation. Let O denote the origin of the Cartesian coordinate system Oxyz fixed to the non-moving segment and let $\Omega$ be the origin of the coordinate system $\Omega\xi\eta\zeta$ which is fixed to the moving segment. Both segments are, of course, connected by the joint. The situation is shown in Fig. 7.2.

Let the j-th muscle spanning the joint have its origin at point D and its insertion on the moving segment at point B. The point D is fixed in

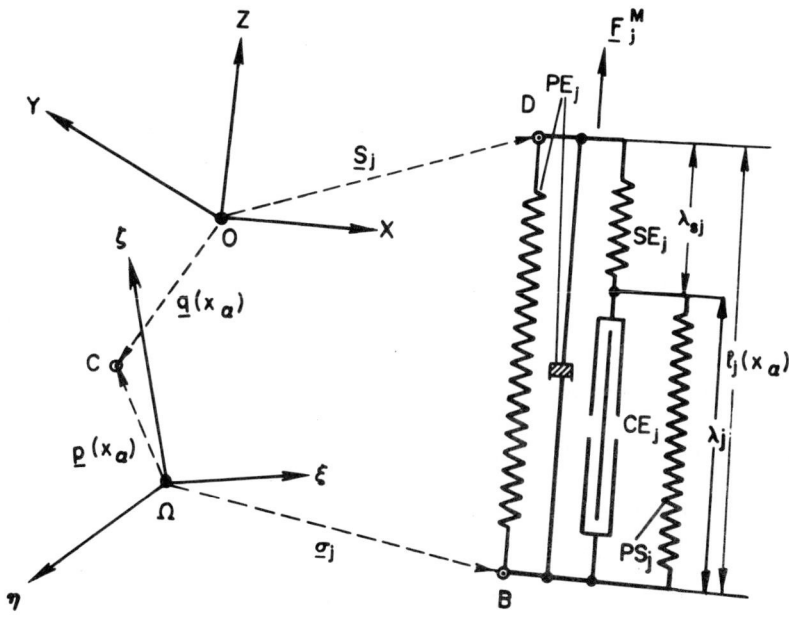

**FIG. 7.2:** Diagrammatic representation of two body segments connected by a joint with a non-constant centre of rotation C. The joint is spanned by a muscle with origin at point D and insertion at point B. Symbols are explained in the text. Note that the cross-bridge elastic element $BE_j$ has been absorbed into the element $CE_j$, which is possible by virtue of relation (4.45)

the system Oxyz and is given by the vector $s_j$ while B is a fixed point in $\Omega\xi\eta\zeta$ and given by $\sigma_j$. The vector $F_j^M$ is the force vector of the j-th muscle measured relative to $\Omega\xi\eta\zeta$. Finally, the vectors $q(x_\alpha)$ and $p(x_\alpha)$ denote the position of the instantaneous centre of rotation C relative to the frames Oxyz and $\Omega\xi\eta\zeta$ respectively. As indicated, these vectors are functions of the three Eulerian angles $x_\alpha$ which describe the orientation of the system $\Omega\xi\eta\zeta$ relative to the system Oxyz.

The model of the fusiform muscle displayed in Fig. 7.2 is identical with that shown in Fig. 4.3, except for the element BE (see explanation in the legend of Fig. 7.2). If a number of $w_i$ muscles span the i-th joint then it is easily shown that the following relations (details in Hatze, 1977b) hold:

$$\mathbf{F}_{ij}^M = [F_{ij}^{PE}(\ell_{ij}(x_\alpha),\dot{\ell}_{ij}(x_\alpha)) + F_{ij}^{SE}(\ell_{ij}(x_\alpha),\lambda_{ij})] \mathbf{f}_{ij}, \qquad (7.12)$$

$$\mathbf{N}_i^M = \sum_{j=1}^{w_i} [\sigma_{ij} - \mathbf{p}_i(x_\alpha)] \times \mathbf{F}_{ij}^M, \qquad (7.13)$$

$$\begin{bmatrix} Q_{\psi i}^M \\ Q_{\theta i}^M \\ Q_{\phi i}^M \end{bmatrix} = \begin{bmatrix} \sin\theta_i\sin\phi_i & \sin\theta_i\cos\phi_i & \cos\theta_i \\ \cos\phi_i & -\sin\phi_i & 0 \\ 0 & 0 & 1 \end{bmatrix} \begin{bmatrix} N_{i\xi}^M \\ N_{i\eta}^M \\ N_{i\zeta}^M \end{bmatrix} \qquad (7.14)$$

where $F_{ij}^M$ is the total (active and passive) force output of the j-th muscle spanning the i-th joint, $\mathbf{f}_{ij}$ is a unit vector in the direction of $\mathbf{F}_{ij}^M$, $\mathbf{N}_i^M$ is the total torque vector (relative $\Omega\xi\eta\zeta$) produced by the $w_i$ myoactuator torques of the muscles across this joint, and $Q_{\psi i}^M$, $Q_{\theta i}^M$, $Q_{\phi i}^M$ denote the generalized myoactuator forces (moments) corresponding to Eulerian angles $\psi_i$, $\theta_i$, $\phi_i$, being the $x_\alpha$ for this joint. Note that $F_{ij}^{PE}$ is to be included in (7.12) *only* if it has not been accounted for in the passive joint moment vector $Q^L$, and that the transformation (7.14) is necessary since the components of $N_i^M$ are directed along the axes $\Omega\xi\eta\zeta$, while the components of $Q_i^M$ are directed along the respective angular velocity vectors of the Eulerian angles and these vectors do not coincide with the axes $\Omega\xi\eta\zeta$. Relations that are similar to (7.12–7.14) can also be derived for muscles spanning more than one joint.

Since $F_{ij}^{SE}$ appears (via $\ell_{ij}(x_\alpha)$ and $\lambda_{ij}$) in the last of Equations (4.37), the interrelation between the executor and the myoactuator subsystems becomes obvious.

Let there be a number m of myoactuators (muscles) acting on the assemblage of limb segments of the hominoid. Then there are m sets of Equations (4.37) constituting the *myoactuator subsystem* of the total model. By defining for each of the m myoactuators the combined state

$$\mu_k \triangleq (n_k, r_k, \psi_k, \varphi_k, \xi_k)^T, \quad k=1,\ldots,m, \quad (7.15)$$

the system relations (4.37) can be written concisely as

$$\dot{\mu}_k = g_k(x, \mu_k, v_k(t), z_k(t)), \quad \mu_k(0) = \mu_k^o, \quad (7.16)$$

or, in vector form, as

$$\dot{\mu} = g(x, \mu, v, (t), z(t)), \quad \mu(0) = \mu_o, \quad (7.17)$$

where $v(t)$ and $z(t)$, $t \in [t_o, t_f]$, are the m-dimensional neural control (input) vector functions.

### 7.1.4 The model of the complete neuro-musculoskeletal system

It is obvious from (7.12–7.14) that the vector function $Q^M$ in (7.9) should be written as $Q^M = Q^M(x, \mu)$ since it depends on both the state $x$ of the executor subsystem and the state $\mu$ (via $\lambda_{ij}$ in (7.12)) of the myoactuator subsystem. Hence the generalized force vector $Q$ appearing in (7.6), and defined by (7.9), must be written as $Q = Q(x, \mu, t)$. The systems (7.6) and (7.17) may now be combined to yield the model of the complete neuro-musculoskeletal control system represented by the non-autonomous, nonlinear, ordinary first-order differential system

$$\dot{x} = A^{-1}(x)[B(x) + Q(x, \mu, t)], \quad x(0) = x_o,$$
$$\dot{\mu} = g(x, \mu, v(t), z(t)), \quad \mu(0) = \mu_o. \quad (7.18)$$

In this model the control vector functions $v(t)$ and $z(t)$ are specified as given input functions of time. They may, however, also result from the application of some optimization procedure, in which case they are optimal controls with respect to certain performance criteria (see Section 7.2). Alternatively, these controls may be the outputs of a model of the neural (controller) subsystem, which could be appended to (7.18), and may contain feedback loops. A detailed discussion of these possibilities is, however, far beyond the scope of this book and will therefore not be pursued further.

### 7.1.5 Simulation of the long-jump take-off phase

As an application to a real-life problem of model (7.18) we shall describe the simulation of the long-jump take-off phase of a given athlete. Before this is done, some aspects of the practical implementation of the model will be discussed.

It is a formidable task to develop an algorithm permitting the execution on the computer of the differential system (7.18) and its associated subprocedures. A total of some 2300 man-hours was required to complete this project. This heavy investment is due to the extreme complexity of the detailed form of the matrix A(x) and the vector B(x), and the multitude of numerical problems that arose during the design and testing of special integration routines and algorithms to enable the right-hand sides of (7.18) to be efficiently computed.

The computer program ANDYMO, the end-product of this effort, is written in ANSI FORTRAN IV and exhibits the following features. Given the set {P} of subject-specific input parameters, the initial state $(x_o,\mu_o)$, and neural control vectors $v(t)$, $z(t)$, $t \in [t_o,t_f]$, the program computes: the state trajectory $(x(t),\mu(t))$ which includes the visible motion $q(t)$ of the hominoid; the histories of all components of the constraint forces of activated constraints at the heels and foot-balls of both feet, at the hands (the model permits handstands, exercises on the horizontal bar, etc.), and at the head (e.g. during a head stand); the trajectory of the centre of mass of the hominoid; the histories of the components of the centre-of-mass velocity; the history of the total angular momentum; the histories of all muscle and joint reaction forces; and the histories of the individual mechanical energies and powers of all the 17 segments of the hominoid shown in Fig. 7.1.

It is noteworthy that the model and its associated computer program can also be used most advantageously to find the forces and torques which produced a certain observed motion, i.e. to solve the inverse dynamic problem. Assume that a configurational trajectory $q(t)$, $t \in [t_o,t_f]$, has been observed as a noise-contaminated data sequence vector $\{q_i(t_k), i=1,\ldots,f, k=0,1,\ldots,N\}$. Then optimally filtered Fourier approximation (Hatze, 1980e) can be used to estimate rapidly and efficiently the first and second derivatives $\dot{q}(t)$, $\ddot{q}(t)$. By definition, $x(t) = (q(t),\dot{q}(t))^T$, so that $x(t)$ and $\dot{x}(t)$ are known functions of time.

Considering only the lower half of (7.8) we find from (7.6) that

$$\bar{Q}(t) = C(x(t))\dot{\bar{x}}(t) - \bar{B}(x(t)), \quad t_o \leq t \leq t_f, \qquad (7.19)$$

where $\dot{\bar{x}}(t) \triangleq \ddot{q}(t)$, $\bar{Q}(t) = \bar{Q}^M(t) + \bar{Q}^L(t) + \bar{Q}^E(t) + \bar{Q}^C(t)$, and

$$C = \left[\frac{\partial^2 L}{\partial \dot{q}_i \partial \dot{q}_j}\right], \quad i,j=1,\ldots,f.$$

In the absence of further information or additional assumptions it is obviously impossible to decompose $\bar{Q}(t)$ into its respective constituents (a detailed discussion of this problem can be found in Hatze, 1980b). However, if $\bar{Q}^E(t) \equiv 0$, which is the case in most applications, then by means of (7.19) it is possible to estimate $(\bar{Q}^M(t)+\bar{Q}^L(t))$ fairly accurately if only one constraint is active. The reason is that $\bar{Q}^M$ and $\bar{Q}^L$ are moments, and hence do not appear as applied forces in the equations for the three linear configurational coordinates $x = q_1$, $y = q_2$, and $z =$

$q_3$. Hence, for three-dimensional motions, the first three components of $\bar{Q}(t)$ are $\bar{Q}_1^C(t)$, $\bar{Q}_2^C(t)$, and $\bar{Q}_3^C(t)$, which can thus be directly computed from (7.19) during constraint activation. Since the constraint torque components $\bar{Q}_j^C(t)$, $j=4,\ldots,f$, depend on $\bar{Q}_i^C(t)$, $i=1,2,3$, the former can also be evaluated, thus permitting the moment vector $(\bar{Q}^M(t)+\bar{Q}^L(t))$ to be estimated from (7.19).

If more than one constraint is active (e.g. heel and ball of a foot on the ground) this procedure becomes invalid, and other methods must be applied. Attempts to decompose the components $\bar{Q}_j^M(t)$, $j=4,\ldots,f$, of the myoactuator moment vector $\bar{Q}^M(t)$ into their respective constituents (if more than one muscle acts across a joint) lead to the well-known myodynamic indeterminacy problem (fully discussed in Hatze, 1980b).

It goes almost without saying that the computer program executing the model can also be used to compute, for an observed motion $\{q(t_k), k=0,1\ldots,N\}$, the instantaneous positions of the body centre of mass, its instantaneous velocity components, the instantaneous total angular momentum, the instantaneous energies (kinetic and potential) and mechanical powers of all the 17 model segments, for each discrete time instant $t_k$. This feature of the model together with the above-mentioned optimal Fourier filtering of the data, makes it possible rapidly to process and evaluate motion data acquired by film analysis or with the SELSPOT system.

As mentioned above, the program ANDYMO requires as input the numerical values of the subject-specific parameter set $\{P\}$. This set consists of the subsets of the segmental, articular, morphometric, myodynamic and myocybernetic parameters which are characteristic of a given individual.

The segmental input parameters are, for each of the 17 segments: the principal moments of inertia, the mass, the coordinates of the mass centroid, and the coordinates of the segment origin. A new anthropometric-computational method (Hatze, 1980d) and the associated computer program SEMCI were used to determine this parameter set for the given athlete. The method is based on a battery of 242 anthropometric measurements taken directly from the subject, and makes it possible to determine all parameter values with an average error of about 2.7% (max. error 5%). The computation of the whole parameter set by the program SEMCI takes 0.515 seconds on a CDC CYBER 174 digital computer.

The set of articular parameters defines the subject-specific vector $Q^L(x)$. This vector accounts for the highly nonlinear passive elasticity and viscosity in the joints. Methods of determining the parameter values entering the components of this vector are discussed in Hayes and Hatze (1977), and Hatze (1975b). Parameters defining varying muscle-moment arms due to nonstationary joint centres of rotation also belong to the articular parameter set.

Finally, the sets of morphometric, myodynamic and myocybernetic parameter values for all muscle groups in the model are determined by the techniques described in Chapter 5.

The 46 muscles in the model are the following, the number of the muscle being given in brackets: splenius capitis and semispinalis capitis (1), left levator scapulae and left upper part of trapezius (2), right parts of these muscles (3), sternothyroid and omohyoid (4), left rhomboideus and left lower part of trapezius (5), right parts of these muscles (6), left pectoralis minor and serratus anterior (7), right parts of these muscles (8), left (9) and right (10) teres major, left (11) and right (12) triceps long head, left (13) and right (14) latissimus dorsi, left (15) and right (16) pectoralis major, left (17) and right (18) biceps brachii, left (19) and right (20) triceps short head, left (21) and right (22) brachialis and brachioradialis, left (23) and right (24) hand flexors, left (25) and right (26) hand extensors, multifidus plus erector spinae and quadratus lumborum (27), obliquus externus and rectus abdominis (28), left (29) and right (30) gluteus maximus, left (31) and right (32) ilio-psoas, left (33) and right (34) hamstring group, left (35) and right (36) popliteus and biceps femoris short head, left (37) and right (38) vasti group, left (39) and right (40) rectus femoris, left (41) and right (42) gastrocnemius, left (43) and right (44) soleus, and left (45) and right (46) anterior leg muscles. For three-dimensional simulations additional leg muscle groups must be included (described elsewhere).

We shall now discuss the simulation of the long-jump take-off. The above set of parameter values was determined for a male athlete aged 24 years. The set of initial values $x_o$, $\mu_o$ in (7.18) was estimated from film analyses of long jumps, and observed corresponding muscle action sequences. The neural control functions $v_i(t)$, $z_i(t)$, for all i=1,...,46 muscle groups were initially guessed from electromyographic records of comparable locomotory movements (Brandell, 1973, and others) and then successively adjusted to produce the required trajectory.

**FIG. 7.3:** 17-segment hominoid during simulated long-jump take-off. Time interval between frames is 0.04 seconds. Ground contact time is approximately 0.110 s, and lift-off angle is 0.3456 rad (19.8 degrees)

133

With these inputs, the differential system (7.18) was simulated for a time interval of 0.16 seconds. The resulting configurational trajectory $\{q_i(t), i=1,\ldots,21\}$, for planar motion, was stored on tape and used to construct a kinematic sequence displaying on a Tektronix 4014 Graphic Display Unit the long-jump take-off motion of the hominoid.

Five sequential frames of that motion, at time intervals of 0.04 seconds, are shown in Fig. 7.3. This figure also clearly shows the appearance of the hominoid consisting of the following 17 segments: abdomino-thoracic segment, head-neck segment, left and right shoulders, upper arms, lower arms, and hands, abdomino-pelvic segment, left and right thighs, legs, and feet.

An in-depth discussion of the many interesting details of this simulation (specific form of the neural controls v(t) and z(t), the trajectory of the centre of mass of the hominoid, the history of the total angular momentum, the sensitivity to perturbations of initial values and other parameters of take-off angle and take-off velocity components, the history of the ground reaction forces, etc.) is clearly beyond the scope of this monograph. Some of these features have been reported elsewhere (Hatze, 1980c) and the reader is therefore referred to this reference. Nevertheless, it is evident that the global myocybernetic muscle model developed in the previous chapters can be used successfully as myoactuator in extremely complicated models of the total human neuromusculoskeletal system.

To conclude this section, a brief exposition of the *optimization problem of the long jump* will be given. Instead of guessing initial values $(x_o,\mu_o)$, and controls v(t), z(t), $t \in [t_o,t_f]$, to produce an arbitrary trajectory $(x(t),\mu(t))$, (i.e. an arbitrary take-off motion of the hominoid), one may attempt to find that set

$$\{\hat{x}_o,\hat{\mu}_o,\hat{v}(t),\hat{z}(t),\hat{h}_e; t \in [0,t_f]\} \quad (7.20)$$

which, for a given individual characterized by the parameter set $\{P\}$, generates a motion which maximizes the distance jumped. In (7.20), $t_f$ denotes the time of the take-off motion, and $h_e$ is explained below.

It is obviously necessary to know the objective function, i.e. the function to be maximized. It can be shown that his function is given by

$$d(x(t_f),h_e) = \rho_{yf} + \dot{\rho}_{yf}\dot{\rho}_{zf}/g + [\ |H_f|/M + \dot{\rho}_{yf}(\dot{\rho}_{zf}^2/g + 2\rho_{zf} - h_e)].$$
$$\cdot [\dot{\rho}_{zf}^2 + 2g(\rho_{zf} - h_e)]^{-\frac{1}{2}}, \quad \rho_{zf} \geq h_e > 0, \quad (7.21)$$

where $d(\cdot)$ is the distance jumped; $h_e$ is the vertical distance from ground level of the centre of mass of the hominoid at the instance of landing; $\rho_{yf}$, $\rho_{zf}$ denote respectively the horizontal and vertical distances of the centre of mass from the front edge of the take-off board at the moment of lift-off; $\dot{\rho}_{yf}$, $\dot{\rho}_{zf}$ are the corresponding horizontal and vertical velocity components; g is the gravitational constant; and M and $|H_f|$ denote respectively the total mass of the athlete and the magnitude of the total angular momentum vector at lift-off.

Expression (7.21) takes into account that the negative angular momentum (forward rotation) must be sufficient to prevent a falling back, for a given landing position. Obviously, this stipulation still leaves the choice of $h_e$ free.

It is clear that all entries in (7.21), except $h_e$, g and M, depend on the state $x(t_f)$ of the athlete at the moment of lift-off, i.e. on the positions and generalized velocities of all his segments at that instant. However, the lift-off state $x(t_f)$ is generated by the controlled take-off motion and the initial state $(x_o, \mu_o)$ in which the take-off motion began. Hence the optimization problem of the long jump can be formulated as follows.

For a given parameter set $\{P\}$ (i.e. for a given athlete) find an optimal initial state $(\hat{x}_o, \hat{\mu}_o)$, optimal controls $(\hat{v}(t), \hat{z}(t), t \in [t_o, t_f])$, and an optimal value $\hat{h}_e$, that maximize (7.21), subject to the differential constraint (7.18) and certain other constraints. Expressed differently: we seek those initial configurations, limb velocities, and excitative and contractive states of all the muscles of the athlete, and those neural control functions transferring his body from the initial take-off state to the final lift-off state, that produce a lift-off state that maximizes the distance jumped, subject to an optimal landing position characterized by the parameter $h_e$.

The techniques and problems associated with the optimization of the long jump are rather involved owing to the high dimensionality and nonlinearity of the model, and the methods are still at a developmental stage. However, results obtained so far are very encouraging and indicate that the present approach may provide us with a powerful technique for optimizing sports motions. In this connection it should be noted that our approach is fundamentally different from that of Ballreich (1973), Luhtanen and Komi (1979), and others. In these works, no formulation of the objective function and the optimization problem is attempted. The treatment is empirical and directed towards identifying measurable kinematic and kinetic quantities or 'features' (Ballreich, 1973), such as step length and step frequency in the run-up, take-off time, components of linear segmental impulses (Luhtanen and Komi, 1979), etc., which are believed significantly to influence the performance of the long-jumper. The selection of these quantities is not based on a conceptual model but on intuition. As is easily demonstrated, these quantities are, in fact, composites of the fundamental parameter set

$$\{x_o, \mu_o, v(t_k), z(t_k), x(t_f), h_e; k=0,\ldots,N\},$$

i.e. they are mutually interdependent and hence quite unsuitable as performance indicators. In fact, without a concise conceptual model, including a performance criterion of the long jump, it is impossible to say which of the parameters labelled as significant are truly independent and hence relevant to the optimization problem under consideration. This is clearly demonstrated by the complexity of the dynamics of the system (7.18) and the objective function (7.21).

## 7.2 FOUNDATIONS OF MYOCYBERNETICS

The optimality principles underlying neuromuscular control have until very recently received comparatively little attention from the biocybernetist, although their investigation represents an extremely interesting and rewarding field of research. This apparent lack of interest is surprising in two respects. First, there exists a vast literature on the phenomenology of neuromuscular behaviour, based on data collected by a great many experimental biologists. Secondly, control theory has advanced to such an extent that it has become possible to treat, from a control point of view, highly complex and nonlinear biosystems subject to severe constraints.

Possibly workers have been discouraged by the lack of an appropriate control model of (skeletal) muscle that would contain the equivalents to the actual neural controls, firing frequency and number of motor units recruited, and be capable of predicting the vast variety of known phenomena of muscular contraction.

This obstacle has now been overcome and it appears that the rich opportunities existing in the emerging discipline of *myocybernetics* can now be fully exploited.

It is not the purpose of this book to present a detailed account of the literature in this field that has appeared up to now. Rather, an exposition will be given of the foundations of myocybernetics. This will enable the reader to gain a basic understanding of this subject and, it is hoped, stimulate further research in this absorbing field.

Since the energetics of muscular contraction plays an important role in myocybernetics, we begin by deriving an expression for the rate of expenditure of energy during various modes of contraction.

### 7.2.1 Myoenergetics

The total energy rate $\dot{E}$ of a muscular contraction can be expressed (Mommaerts, 1969) as the sum of the activation heat rate $\dot{g}$, the maintenance heat rate $\dot{h}$, the shortening heat rate $\dot{s}$, the work rate $\dot{w}$, and the rate $\dot{p}$ of heat dissipated in the parallel elastic structures. Hence

$$\dot{E} = \dot{g} + \dot{h} + \dot{s} + \dot{w} + \dot{p}. \qquad (7.22)$$

Note that the symbol $\bar{h}$ designating the maintenance heat rate is *not* synonymous with $h(\dot\xi)$, the symbol for the velocity-dependence function as defined by (4.15).

The *activation heat* is thought to be connected to the release and re-uptake of Ca by the sarcoplasmic reticulum, and is that portion of the total internal heat production which is not connected to the actomyosin system (Woledge, 1971). The activation heat per stimulus has been measured in slow and fast rat muscle (Gibbs and Gibson, 1972; Wendt and Gibbs, 1973), and was found to be an exponential function of the stimulus interval $v^{-1}$. Multiplying this function by the stimulation rate $v$ we obtain the activation heat rate $\dot{g}$ for a muscle having a mass of G kg of which a proportion u is active as

$$\dot{g} = uG\bar{g}v[1-\exp(-\kappa_8-\kappa_7/v)]/[\hat{v}-\hat{v}\exp(-\kappa_8-\kappa_7/\hat{v})], \qquad (7.23)$$

where $\bar{g}$ (in W/kg muscle mass) is the muscle-specific activation heat rate constant, $\hat{v}$ is the maximum stimulation frequency occuring, and the constants $\kappa_7$ and $\kappa_8$ have values of 18.2 and 0.25 respectively. Wendt and Gibbs (1973) have also determined values of the ratio $\bar\phi$ defined by

$$\bar\phi = \bar{g}/(\bar{g}+\bar{h}), \qquad (7.24)$$

where $\bar{h}$ is the muscle-specific maintenance heat rate constant. The values for $\bar\phi$ were found to be about 0.35 for fast mammalian fibres and 0.45 for slow mammalian fibres (see Table 2 of Wendt and Gibbs, 1973), and agree with observations by Edwards *et al.* (1975) who found an average value of 0.34 (since 66% of the heat rate was due to glycolytic reactions) for the human quadriceps muscle. Data for $(\bar{g}+\bar{h})$ for fast and slow human muscle fibres have also been collected (Bolstad and Ersland, 1975). The respective values are 150 W·kg$^{-1}$ and 24.4 W·kg$^{-1}$. These values as well as their ratio agree remarkably well with the data observed in rat muscle at 27°C (Gibbs and Gibson, 1972; Wendt and Gibbs, 1973). They constitute, however, only average values of a population of a certain fibre type, and fairly large deviations from average properties within such populations have been documented (Burke *et al.*, 1971).

Hill (1938) has demonstrated that the combined activation and maintenance heat rates can be expressed as

$$\dot{g} + \bar{h} = ab = (a/P_o)^2 P_o V_o, \qquad (7.25)$$

where it has been found (Ritchie and Wilkie, 1958; Jewell and Wilkie, 1960; Bahler *et al.*, 1968) that Hill's 'constants' $a, P_o, V_o$ are, in fact, complicated functions of the relative length $\xi$ and the active state q of the contractile element. The significance of these experimentally observed functions has been discussed in Chapters 3 and 4, where the following expressions were derived:

$$P_o(\xi,q) = k(\xi)q\bar{F}, \quad (7.26)$$

$$V_o(\xi,q) = -\bar{\lambda}_o(-\dot{\eta}_o(\xi,q)), \quad (7.27)$$

and where $\dot{\eta}_o(\cdot)$ is given by (3.46). For a muscle stimulated artificially as a whole we have, of course, that $\varepsilon \equiv q$, i.e. the excitation is identical with the active state.

It has been found that the ratio $a/P_o$ appearing in (7.25) is fairly independent of the length $\xi$ but depends strongly on the active state q. Indeed, Julian (1971, Figs. 8 and 10) has convincingly demonstrated that at $q=1$ (maximum Ca concentration) $a/P_o$ had a value of 0.18 for the muscle he investigated, while for $q=0.36$ this value increased to 0.24. Since $a/P_o$ is independent of $\xi$, but dependent on q, we can by virtue of the relation (7.26) write

$$a/P_o = k(\xi)\bar{a}q^m/(k(\xi)q\bar{F}) = (\bar{a}/\bar{F})q^{-0.28}, \quad (7.28)$$

where the approximate value of $m=0.72$ has been determined from the above data, and $\bar{a}=a$ for $q=1$.

It must, however, be mentioned that the function (7.28) cannot be viewed with complete confidence. First, the assumption that $a/P_o$ is an exponential function of q can only vaguely be justified by a known similar dependence of $V_o$ on q (Jewell and Wilkie, 1960), and by the fact that $a/P_o = b/V_o$. Secondly, it is always a hazardous undertaking to try to extrapolate data obtained from one type of muscle at a certain temperature to another type working at a different temperature. Some reassurance can, however, be derived from the fact that the dependence of $V_o$ on q as found in frog muscle at $0°$ C is qualitatively similar to the dependence expected for mammalian muscle at $37°$ C.

Substituting (7.26), (7.27), and (7.28) into (7.25) we obtain

$$\dot{g} + \dot{h} = k(\xi)q^{0.44}\bar{F}(\bar{a}/\bar{F})^2\bar{\lambda}_o\dot{\eta}_o. \quad (7.29)$$

It is remarkable that the frequently observed length-dependence of the heat rate $\dot{g} + \dot{h}$ (Abbott, 1951; Gibbs and Gibson, 1972) is fairly accurately predicted by the function (7.29). Since Equation (7.29) gives the heat rates for a specific muscle having a mass of G kg we must have

$$-(\bar{a}/\bar{F})^2\bar{F}\bar{\lambda}_o = (\bar{\dot{g}}+\bar{\dot{h}})G, \quad (7.30)$$

and if only a proportion u of the muscle is active then (7.29) becomes

$$\dot{g} + \dot{h} = uG(\bar{\dot{g}}+\bar{\dot{h}})q^{0.44}k(\xi)(-\dot{\eta}_o). \quad (7.31)$$

Now, $\dot{g}$ is that part of expression (7.31) which is independent of $\xi$ and constitutes a proportion $\bar{\phi}$ of the total [see (7.24)] when $\xi=1$ and $v=\hat{v}$. We therefore have to reduce $k(\xi)$ by $\bar{\phi}$ and then add $\dot{g}$ to the reduced term. The resulting expression is then

$$\dot{g} + \dot{h} = uG(\bar{\dot{g}}+\bar{\dot{h}})[q(k(\xi)-\bar{\phi})+\bar{\phi}v\,\frac{1-\exp(-\kappa_8-\kappa_7/v\hat{v})}{1-\exp(-\kappa_8-\kappa_7/\hat{v})}], \quad (7.32)$$

where the normalized stimulation rate $v$ is defined by $v=v/\hat{v}$, and use has been made of the fact that $\dot{\eta}_o(\xi,q)$ can be approximated by $-\dot{\eta}_o(\xi,q) \approx q^{0.56}$ (see Fig. 3.7). Note that (7.32) is actually an implicit statement about the nature of the function when $\xi \neq 1$ and $v < 1$.

The *shortening heat rate* $\dot{s}$ produced by a muscle is given by Hill (1938) as $\dot{s}=aV$, where a is the same constant which appears in (7.25) and V is the absolute value of the shortening velocity. If a proportion u of the muscle is active, and if the expression (7.28) for a is used with $m \approx 1$, the shortening heat rate is obtained as

$$\dot{s} = -u\bar{a}\bar{\lambda}k(\xi)q\dot{\xi}, \quad (7.33)$$

where we have put $V=-\bar{\lambda}\dot{\xi}$ (shortening velocities are negative in the present model).

Finally, the *work rate* $\dot{w}$ is obviously represented by

$$\dot{w} = -\bar{\lambda}\dot{\xi}F^{SE}, \quad (7.34)$$

while the *heat rate* $\dot{p}$ dissipated in the viscous component of the parallel elastic element is given by

$$\dot{p} = c_v \dot{\ell}^2, \quad (7.35)$$

where $c_v$ is a constant and $\dot{\ell}$ the velocity of movement of the total muscle, including the series-elastic element.

It would be convenient to be able to express both the shortening heat rate $\dot{s}$ and the work rate $\dot{w}$ in terms of the muscle mass G. This can be achieved as follows. Let it be assumed that the contractile part of the muscle under consideration has been shaped into a rectangular parallelepiped with sides $\ell$, b, and e. Since the density of muscle tissue is approximately 1000 kg.m$^{-3}$, the mass of the purely contractile part of the muscle is given by

$$G = 1000\ell be. \quad (7.36)$$

As is easily shown, the width b for bipennate muscles is defined by

$$b = 2\bar{\lambda}\cos\bar{\theta}, \quad (7.37)$$

where the notation of Chapter 4 has been used. In addition, it can be demonstrated (Hatze, 1980a) that the physiological cross-sectional area A of a bipennate muscle is given by

$$A = 2\ell e\cos\bar{\theta}, \quad (7.38)$$

while for fusiform muscles (for which $\ell \approx \bar{\lambda}$)

$$A = be. \quad (7.39)$$

Using (7.36) and (7.37) in (7.38) and (7.39) we find that for both penniform and fusiform muscles

$$A = G/(1000\bar{\lambda}). \qquad (7.40)$$

Let $\bar{f}$ (in $N \cdot m^{-2}$) denote the maximum isometric force of the muscle per unit of cross-sectional area. Then the maximum isometric force $\bar{F}$ is given by

$$\bar{F} = \bar{f}A = \bar{f}G/(1000\bar{\lambda}), \qquad (7.41)$$

where (7.40) has been used.

Since contractile energy is expended by both the stimulated and semi-active motor unit populations, we must interpret the product uq appearing in (7.32) and (7.33) as being approximately equal to the total excitation ε. On the other hand, activation heat is produced only by stimulated motor units, which implies that u in (7.23) must be defined as

$$u = [\exp(\bar{c}n)-1]/[\exp(\bar{c})-1] . \qquad (7.42)$$

Using these arguments and (7.41) in (7.33) and (7.34), the latter two equations can be written as

$$\dot{s} = -\dot{\xi}G\bar{f}(\bar{a}/\bar{F})\epsilon k(\xi)/1000 , \qquad (7.43)$$

and

$$\dot{w} = -\dot{\xi}G\bar{f}F^{SE}/(1000\bar{F}) , \qquad (7.44)$$

where $\bar{F}$ is always measured in the direction of the muscle fibres, and $F^{SE}$ must be replaced by $F^{ST}/\sin\theta$ for penniform muscles (see Section 4.4).

Combining (7.32), (7.42), (7.43), (7.44), and (7.35), Equation (7.22) becomes

$$\begin{aligned}\dot{E} = \ & G\{(\bar{\bar{g}}+\bar{h})[\epsilon(k(\xi)-\dot{\phi}) + \\ & + \frac{\exp(\bar{c}n)-1}{\exp(\bar{c})-1} \dot{\phi}v \frac{1-\exp(-\kappa_8-\kappa_7/v\hat{v})}{1-\exp(-\kappa_8-\kappa_7/\hat{v})} ] - \\ & - \dot{\xi} \frac{\bar{f}}{1000} [(\bar{a}/\bar{F})\epsilon k(\xi)+F^{SE}/\bar{F}]\} + c_v\dot{\ell}^2, \end{aligned} \qquad (7.45)$$

where ε and $k(\xi)$ denote respectively the excitation and filamentary-overlap functions, the contractile velocity $\dot{\xi}$ is given by the last of Equations (4.37), and all other symbols have the meanings defined previously. Note that if $(k(\xi)-\dot{\phi})<0$ it must be set equal to zero. The constants with fixed values appearing in (7.45) have been collected in Table 7.1.

TABLE 7.1: Values of muscle-specific constants appearing in (7.45). Units are W·kg⁻¹ for ($\bar{g}+\bar{h}$) and N·m⁻² for $\bar{f}$

|  | $\bar{g}+\bar{h}$ | $\bar{\phi}$ | $\kappa_7$ | $\kappa_8$ | $\bar{f}$ | $\bar{a}/\bar{F}$ |
|---|---|---|---|---|---|---|
| fast muscle | 150.0 | 0.35 | 18.2 | 0.25 | $\approx 10^6$ | 0.28 |
| slow muscle | 24.4 | 0.45 | 18.2 | 0.25 | $\approx 10^6$ | 0.16 |

Relation (7.45) is an *empirical* expression for the rate of *metabolic* energy expended by a contracting muscle. The arguments of the function are the states μ and x (via $\xi, n, \varepsilon, F^{SE}(\ell(x), \xi)$, and $\dot{\ell}(x)$) of the NMSCS, and the control v of the muscle in question. Hence for the i-th myoactuator in the system, the total energy rate function can be written as

$$\dot{E}_i = \dot{E}_i(x, \mu_i, v_i). \qquad (7.46)$$

### 7.2.2 Formulation of the general myocybernetic performance criterion and the myocybernetic optimal control problem

For a large class of muscular contractive modes the neural control patterns are of stereotyped form, suggesting that their configuration is governed by some underlying principles. It has been conjectured by several authors that these principles are, in fact, optimality principles.

Milsum (1971) states that 'Many neuro-muscular systems, including those for locomotion, are capable of multi-modal operation and each of these modes may have its own optimal condition of velocity, in the case of locomotion. A more difficult conceptual consideration now involves how these different optima are weighted into one commensurable cost function for the animal as far as an overall evolutionary strategy is concerned'. The author then conjectures that the minimization of metabolic energy expended and the (obviously non-simultaneous) maximization of the muscular power output probably constitute two major processes involved in the survival strategies of the animal.

Tomović and Bellman (1970) assume the presence, in the nervous system, of a multilevel control structure and postulate the following optimization criterion: 'Among the set of feasible algorithms (control strategies) the one with the maximum number of loose joint states is optimal'. The authors claim that this is equivalent to saying that skeletal activity with minimum (metabolic?) energy is preferred; they do not, however, substantiate their claim.

A valuable contribution has recently been made by Hardt (1978), who evaluated optimization methods used to determine muscle forces in the leg during normal locomotion. Hardt found that the commonly adopted technique of using linear programming methods for resolving the problem of indeterminacy in the executor-actuator system leads to totally unrealistic muscle-force histories, and that the proper role of the optimality concept is to provide a unifying criterion for decision making in neuro-muscular control.

Some other papers dealing with the optimality aspects of neuro-musculoskeletal control systems are those of Chow and Jacobson (1971), Chao and Rim (1973), FitzHugh (1977), and Hatze (1976).

There have also been a number of experimental studies (Brett (1965), Otis et al. (1950), Cotes and Meade (1960), etc.) which indicate the presence of optimality principles in myocybernetic control. In particular, it becomes apparent that in all likelihood the metabolic energy expended is the most important part of the general multicriterion. We shall elaborate on this point presently.

In order to discover the myocybernetic performance criteria which lead to certain neural control patterns, one would have to solve the corresponding inverse optimal control problem. However, for the complicated and highly nonlinear neuro-musculoskeletal control system such an approach is not feasible. Instead, one assumes, on logical grounds, the presence of certain performance criteria and then determines the corresponding optimal controls. Coincidence of predicted and observed controls would be a strong indication of correct choice of the respective performance index.

The *myocybernetic performance criterion* in its most general form can be presented as

$$J = \int_{t_o}^{t_f} [\sum_{i=1}^{m} \Lambda_i \dot{E}_i(x,\mu_i,v_i) + L_o(x,\mu,v,z,t)] \, dt + \Phi(x(t_f),\mu(t_f),t_f), \quad (7.47)$$

where $\Lambda_i$, $i=1\ldots,m$, are weighing factors; m is the total number of muscles in the system; $\dot{E}_i(\cdot)$ denotes the rate function of metabolic energy expenditure of the i-th muscle given by (7.45); and $L_o(\cdot)$ and $\Phi(\cdot)$ are certain scalar functions of their respective vector arguments. the symbols $t_o$ and $t_f$ denote initial and final time respectively, while all other symbols have the meanings defined previously.

Associated with (7.47) is the *general myocybernetic optimal control problem* which can be formulated as follows.

For given functions $L_o(x,\mu,v,z,t)$, $\Phi(x(t_f),\mu(t_f),t_f)$, $\dot{E}_i(x,\mu_i,v_i)$, and constants $\Lambda_i$, $i=1\ldots,m$, find control vector functions $v(t),z(t)$, $t \in [t_o,t_f]$, and, possibly, initial values $x_o$ and $\mu_o$, that afford a minimum value to

(7.47) subject to the differential constraint (7.18), i.e.

$$\dot{x} = A^{-1}(x)[B(x)+Q(x,\mu,t)], \quad x(0) = x_o,$$
$$\dot{\mu} = g(x,\mu,v(t),z(t)), \quad \mu(0) = \mu_o,$$

the control vector constraints

$$-\check{z} \leq z \leq 1, \quad 0 \leq v \leq 1; \tag{7.48}$$

the myo-state inequality constraint

$$0 \leq n_i \leq 1, \quad i=1,\ldots,m; \tag{7.49}$$

possible state equality constraints of the form

$$S(x,\mu,t) = 0, \tag{7.50}$$

and possible terminal constraints of the type

$$G(x(t_f), \mu(t_f), t_f) = 0. \tag{7.51}$$

The final time $t_f$ may be fixed or free.

There are numerous interesting special cases of this problem, some of which will be briefly discussed in the next section.

### 7.2.3 Examples of optimal myocybernetic control modes

If the transition process $(x(t),\mu(t),v(t),z(t))$, $t \in [t_o,t_f]$, of the biosystem is to be *time-optimal,* then $\Lambda_i \equiv 0$, $i=1,\ldots,m$ $L_o(\cdot) \equiv 1$, and $\Phi(\cdot) \equiv 0$ in (7.47). Such a problem was solved successfully and is reported in (Hatze, 1976) where, however, certain terminal constraints were included in $\Phi(\cdot)$, in penalty function form, so that $\Phi(\cdot) \not\equiv 0$ in this case.

In this specific problem it was required to find that kicking motion which transfers the right leg of a given human subject, standing on his left leg with the hip belted to a steel frame, from a vertical position (leg muscles relaxed) to a position where the hip joint is flexed at 0.8 rad and the knee joint extended, in the shortest possible time. A male subject was fitted on his right foot with a special boot carrying a mass of 10 kg (in order to slow down the otherwise too rapid motion). He was told to kick at a target, positioned appropriately in front of him, as fast as possible. As stated above, the performance criterion is determined by putting $\Lambda_i \equiv 0$, $L_o(\cdot) \equiv 1$, and $\Phi(x(t_f)) = (x_{i+1}(t_f)-0.8)^2 + (x_{i+2}(t_f)-0.05)^2 + (x_{i+3}(t_f))^2$. The function $\Phi(\cdot)$ ensures that the endpoint conditions (hip angle at 0.8 rad, knee angle at 0.05 rad, angular knee velocity $x_{i+3}$ equal to zero) are satisfied.

A first-order algorithm of Differential Dynamic Programming was used to obtain the numerical solution of this time-optimal control problem, and this solution was compared with the experimentally observed performance of the subject. It was found that any motion of the subject which deviated from the predicted optimal one took a longer time to complete and hence was not optimal. Moreover, the trajectories and controls of near-optimal motions as measured on the living system were indeed found to be close to the theoretical optimal process, thus again confirming the optimality of the model solution (for details see Hatze, 1976).

An interesting class of problems involves transition processes with *maximum performance* of the biosystem during a short movement cycle (processes such as the take-off phase of the high and long jump). In this case we have $\Lambda_i \equiv 0$, $i=1,\ldots,m, L_o(\cdot) \equiv 0$ and $\Phi(\cdot) \neq 0$. For the long jump, for instance, the performance criterion is the distance, $d(\cdot)$, jumped, which must be maximized (or, equivalently, $-d(\cdot)$ must be minimized). Hence, apart from the parameter $h_e$, which is adjusted during the flight phase (i.e. after $t_f$) and hence independent of the trajectory $x(t)$ for $t \leq t_f$, the function $\Phi(\cdot)$ for this problem is identical with $-d(x(t_f);h_e)$ as given by (7.21). It is then required to find that set (7.20) of initial conditions $x_o$, $\mu_o$, parameter value $h_e$, and control functions $v(t)$ and $z(t)$, which minimizes $\Phi(\cdot)$ (i.e. maximizes the distance d).

Important problems arise when neuromuscular transition processes are to be investigated involving *minimization of metabolic energy expenditure*. In this case, $\Lambda_i \neq 0$, $i=1,\ldots,m$, $L_o(\cdot) \equiv 0$ and $\Phi(\cdot)$ may or may not be zero.

A particularly interesting case is that of *steady-state isometric contraction*. Let it be assumed that the muscle is kept at a fixed length $\ell$ where the force contribution due to the parallel elastic element is negligible. By (4.41), the force output F of the muscle is equal to $F^{SE}$. However, a certain prescribed value of $F^{SE}$ can be produced in many different ways. On the one hand, relatively few motor units could be recruited, all of which fire at maximum rate. On the other hand, a large number of motor units firing at a minimum rate could also produce the same force. In between these extremes, infinitely many combinations are possible of degree of recruitment and stimulation rate to produce a certain constant force output. However, experimental observations (Gydikov and Kosarov, 1973; Milner-Brown, Stein, and Yemm, 1973b; etc.) indicate that these combinations of degree of motor unit recruitment and stimulation rates as well as the sequential order of recruitment of the individual units are not random processes but appear to follow a certain predetermined pattern in all skeletal muscles so far investigated.

We may now hypothesize that this regular pattern is dictated by a minimum-energy principle, i.e., we assume that at each given level of

steady-state isometric force production the number of active motor units and their firing rates are so chosen by the nervous system that the metabolic energy expended by the muscle per unit of time is minimized. It is easily shown (see Hatze and Buys, 1977) that for the case of steady-state isometric contraction the optimal control problem reduces to the nonlinear programming problem of minimizing $\dot{E}(x,v)$, subject to the constraints $\dot{\xi} \equiv 0$, $0 \leq v \leq 1$, and $0 \leq u \leq 1$, where
$$u \triangleq (\exp(\bar{c}n)-1)/(\exp(\bar{c})-1).$$

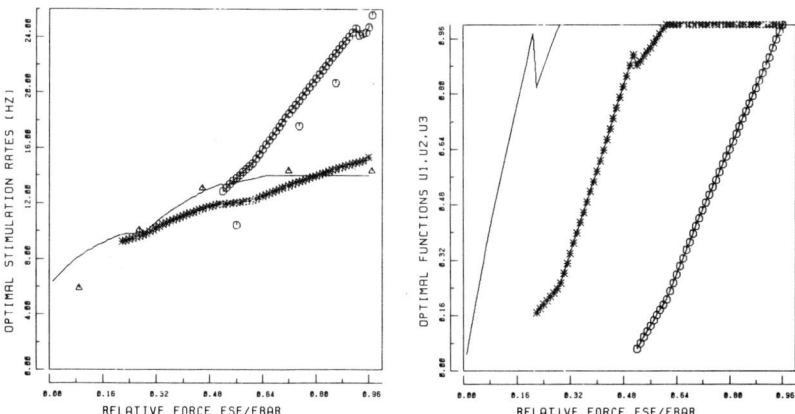

**FIG. 7.4:** *Left-hand part:* predicted energy-optimal stimulation rates for slow (solid line), intermediate (connected stars), and fast motor units (connected circles), and comparable experimental values (single symbols). *Right-hand part:* predicted optimal recruitment patterns for slow (solid line), intermediate (connected stars), and fast units (connected circles). Both graphs relate to the human biceps muscle at optimum length $\bar{\ell}$ (from Hatze and Buys, 1977)

The predicted minimizing stimulation rates and degrees of recruitment for a human biceps muscle comprising three different fibre types (fast, intermediate, slow) are compared with experimental results (single symbols) in Figure 7.4.

From Fig. 7.4 it is obvious that the predicted energy-optimal control patterns agree well with those observed experimentally. Although this fact cannot be taken as a definite proof of energy-optimality in this contractive mode, it nevertheless strongly suggests that such a principle is operative. Indeed, the well-known fact that the smaller and more economical motor units are recruited first and used in muscular endurance tasks requiring the greatest absolute expenditure of metabolic energy, lends further support to the minimum-energy hypothesis. We are thus led to conclude that in all likelihood the specific patterns of motor unit recruitment and firing rate observed in static isometric contractions is determined by a principle of minimum expenditure of metabolic energy.

### 7.2.4 A principle of optimal grading sensitivity

It would appear that the minimization of energy expenditure is not the only teleological adaptation operating in the neuromuscular control system. A type of *structural* adaptation seems to have been realized in what may be called the principle of maximum grading sensitivity.

Replacing N by (N–1) in (4.18) the exact form of this relation is obtained and may be rearranged to yield

$$N = 1 + (\bar{N}/\bar{c})\ln(u/u_o), \quad 0 < u_o \leq u \leq 1, \qquad (7.52)$$

i.e. there exists a logarithmic relationship between the number of motor units recruited and the relative cross-sectional area occupied by these units. Since u is proportional to the relative tetanic force output $\hat{F}/\bar{F}$, it is possible to compare experimental data gathered by Milner-Brown *et al.* (1973b) with the corresponding model prediction defined by (7.52). The result is shown in Figure 7.5, where the dots represent experimental data, while the solid curve shows the model prediction (for

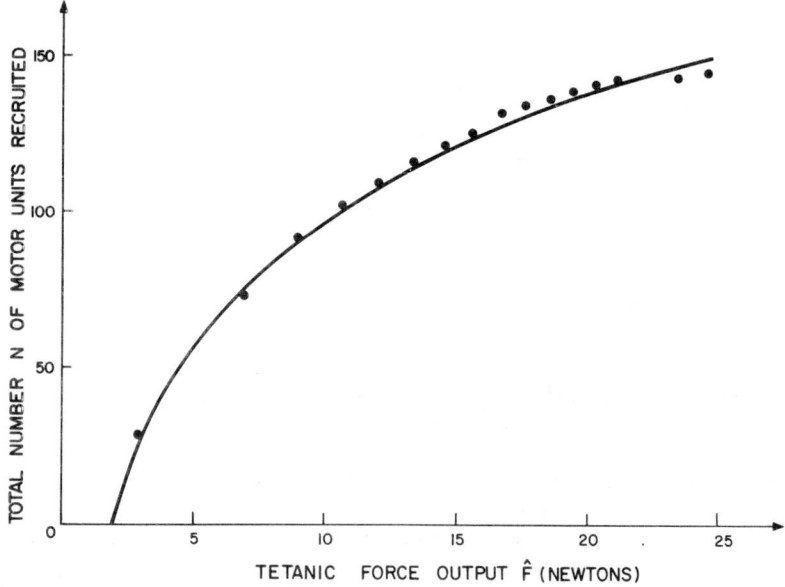

**FIG. 7.5:** Total number N of motor units that are active at tetanic force levels $\hat{F}$. For further explanation see the text (figure adapted from Hatze, 1979)

details consult Hatze, 1979). The good agreement between the two curves is apparent.

It should perhaps be mentioned that $\bar{F}$ (with a value of 24.68 Newton for the muscle whose characteristic is displayed in Figure 7.5) represents, in the present case, *not* the maximum tetanic force of the muscle but only the maximum of the force range investigated. It does, however, correspond to that range, i.e. to the sample of motor units investigated.

It may be of interest to note (see Fig. 7.5) that the agreement between experimental observations and theoretically predicted results is less good at large values of $\hat{F}$. This discrepancy is most probably due to a sampling bias present in the experimental procedure as has been discussed in (Hatze, 1979).

Equation (7.52) may be differentiated with respect to u, and subsequently discretized by replacing dN and du by $\triangle N$ and $\triangle u$ respectively. The result is

$$\triangle u/u = (\bar{c}/\bar{N})\triangle N, \quad (7.53)$$

which, for unit increment of $\triangle N = 1$ (i.e. an increment of one motor unit) resembles Weber's law for the response of a sensory biosystem to the intensity of the input stimulus. In (Hatze, 1979), we were able to demonstrate that Weber's law can be regarded as the realization, by natural adaptation processes, of a *teleological principle of minimum transentropy*.

Now, each motoneurone in the spinal cord is subject to a multitude of neural inputs from higher centres, different feedback fibres, and other motoneurons of the pool in which it is situated. The information transmitted by these input signals is, of course, contaminated by noise. Additional noise is created by the quantal random release of chemical transmitter substance at the inhibitory and excitatory synapses of the motoneurone (Kuffler and Nicholls, 1976). It has been found that even when no input is active, so-called 'spontaneous miniature synaptic potentials' occur, whose amplitudes are statistically normally distributed.

Because the result of the synthesis of all synaptic potentials determines the output firing frequency of a motoneurone, it is obvious that in a network of neurons of vastly different sizes the sequence of neuron activation will be subject to random fluctuations even though an undistorted input would produce the deterministic sequence defined by (7.52). Fluctuations of this kind have indeed been observed experimentally (Gydikov and Kosarov, 1973). We are thus justified in regarding the whole network of motoneurons supplying one muscle as an input-output system that is subject to contamination by noise.

By the similarity of the systems structures (see Hatze, 1979), and by the analogy of the laws describing the input-output relations of sensory biosystems and the sequential order of motor unit recruitment respectively, we are led to conclude that the law described by (7.53) is also a consequence of the minimum-entropy principle. However, since a minimization of the fluctuations in the recruitment sequence implies a maximization of the grading sensitivity of the force output of the muscle, it appears more appropriate to speak of a *principle of maximum grading sensitivity*, the realization of which is the motor unit size law (7.53).

This completes the discussion of the foundations of myocybernetics. It is anticipated that this new discipline will further contribute to the elucidation of the teleological principles governing the behaviour of neuro-musculoskeletal control systems.

# Appendices

**APPENDIX A1**

# Derivation of the state equations of the global myocybernetic control model

By the hypothesis advanced in Chapter 4 there exists a normalized 'average' length $\xi$ of the completely lumped contractile element CE (Fig. 4.3), which length is defined by the contraction dynamics. The latter, in turn, is governed by an excitation function $\varepsilon$ which will be derived below. Using (4.19) and (4.20), Equation (4.34) can be written as

$$F^C/\bar{F} = k(\xi)A[\sum_{i=1}^{N(t)} \exp\{\bar{c}i/\bar{N}\}q(\xi,\gamma_i^+)h_i(\dot{\xi}) +$$

$$+ \sum_{j=N(t)+1}^{N(t)+R(t)} \exp\{\bar{c}j/\bar{N}\}q(\xi,\gamma_j^-)h_j(\dot{\xi}) +$$

$$+ \sum_{k=N(t)+R(t)+1}^{\bar{N}} \exp\{\bar{c}k/\bar{N}\}q_o h_k(\dot{\xi})] - b_1 k_1(\xi). \quad (A1.1)$$

Our objective is now to find functions $\psi, \varphi, h_n, h_r,$ and $h_o$ such that

$$q(\xi,\psi)h_n(\dot{\xi}) \sum_{i=1}^{N(t)} \exp\{\bar{c}i/\bar{N}\} = \sum_{i=1}^{N(t)} \exp\{\bar{c}i/\bar{N}\}q(\xi,\gamma_i^+)h_i(\dot{\xi}), \quad (A1.2)$$

$$q(\xi,\varphi)h_r(\dot{\xi}) \sum_{j=1}^{R(t)} \exp\{\bar{c}(\hat{N}-(j-1))/\bar{N}\} = \sum_{j=1}^{R(t)} \exp\{\bar{c}(\hat{N}-(j-1))/\bar{N}\}q(\xi,\gamma_j^-)h_j(\dot{\xi})$$

$$(A1.3)$$

and
$$q_o h_o(\overset{\shortmid}{\xi}) \sum_{k=N(t)+R(t)+1}^{\bar{N}} \exp\{\bar{c}k/\bar{N}\} = \sum_{k=N(t)+R(t)+1}^{\bar{N}} \exp\{\bar{c}k/\bar{N}\} q_o h_k(\overset{\shortmid}{\xi}), \quad (A1.3a)$$

where we have put

$$\sum_{j=N(t)+1}^{N(t)+R(t)} \exp\{\bar{c}j/\bar{N}\} = \sum_{j=1}^{R(t)} \exp\{\bar{c}(\hat{N}-(j-1))/\bar{N}\},$$

with $\hat{N} = N + R$.

Let the time interval between the recruitment, at times $t_N$ and $t_{N+1}$, say, of the N-th and the (N+1)-st motor units respectively, be $\triangle t(t) \overset{\triangle}{=} t_{N+1}-t_N$. Since the total number $\bar{N}$ of motor units present in a muscle is usually large (hundreds to thousands) we are justified in approximating the step function $N(t)/\bar{N}$, $0 \le t \le T$, by the *absolutely continuous* and *piecewise differentiable* function n(t), defined by

$$\dot{n}(t) \overset{\triangle}{=} \hat{n}z(t) \overset{\triangle}{=} [N(t+\triangle t(t))-N(t)]/\bar{N}\triangle t(t) = 1/\bar{N}\triangle t(t), \quad (A1.4)$$

$$n(0) = 0, \quad 0 \le n \le 1,$$

where $z(t)$, $-\check{z} \le z(t) \le 1$, and hence $\dot{n}(t)$, $t \in [0,T]$, are *piecewise constant* functions of t. For the definition of all the symbols see the main text.

From (A1.4) it follows that

$$\triangle t(t) = 1/[\bar{N}\hat{n}z(t)], \quad (A1.5)$$

where $z(t) \ne 0$ is constant for $t_N \le t \le t_N + \triangle t(t)$, N=1,2,..., and $\hat{n}$ denotes a constant.

Assume that at the time $t_i$ of stimulation there is a rest Ca-concentration $\varphi(t)$ in the i-th motor unit and that the control $v_i(t)$ of this unit is constant for $t \in [t_i, t_i+\triangle_i t)$. Then $\gamma_i^+(v_i(t-t_i))$ appearing in (A1.2) can be found by integrating Equation (3.27):

$$\gamma_i^+(v_i(t-t_i)) = cv_i[1-\exp\{-m_i(t-(i-1)\triangle t(t))\}] + \varphi(t), \quad (A1.6)$$

with $\triangle t(t)$ given by (A1.5), and $m_i$ indicating that m may be different for each unit.

In general, it will not be possible to find functions $\psi(\cdot)$ and $\varphi(\cdot)$ such that (A1.2) and (A1.3) are satisfied exactly for the whole domain of all function arguments. We shall, however, see that a sufficiently good approximation is possible.

Since $\bar{N}$ is generally large, we may use an integral approximation to the sums in (A1.2) and (A1.3), so that we require

$$q(\xi,\psi)h_n(\ddot\xi) \int_1^{\bar N(t)} \exp\{\bar cx/\bar N\}dx = \int_1^{\bar N(t)} \exp\{\bar cx/\bar N\}q(\xi,\gamma^+(x))h(\ddot\xi,x)dx, \quad (A1.7)$$

and

$$q(\xi,\varphi)h_r(\ddot\xi) \int_1^{R(t)} \exp\{\bar c(\hat N+1-x)/\bar N\}dx =$$

$$= \int_1^{R(t)} \exp\{\bar c(\hat N+1-x)/\bar N\}q(\xi,\gamma^-(x))h^*(\ddot\xi,x)dx, \quad (A1.8)$$

where, by virtue of (4.15), (3.36) and (3.37),

$$h_n(\ddot\xi) = \frac{1}{b_2[a_1+\exp\{-a_2\sinh(\ddot\xi a_3 S(n)+a_3/2)\}]}, \quad (A1.9)$$

with $S(n)$ denoting a function (specified below) expressing the dependence on the order number $N$ of a motor unit of $(\vec\lambda/-\vec\lambda_o)$ (see Equation (4.29)),

$$h(\ddot\xi,x) = \frac{1}{b_2[a_1+\exp\{-a_2\sinh(\ddot\xi a_3(a_4-a_5 x/\bar N)+a_3/2)\}]}, \quad (A1.10)$$

since, by (4.29), $(\vec\lambda/-\vec\lambda_o) = a_2'/B - a_3' n/B = a_4 - a_5 N/\bar N$,

$$h_r(\ddot\xi) = \frac{1}{b_2[a_1+\exp\{-a_2\sinh(\ddot\xi a_3 S(n,r)+a_3/2)\}]}, \quad (A1.11)$$

where $S(n,r)$ designates a function similar to $S(n)$,

$$h^*(\ddot\xi,x) = \frac{1}{b_2[a_1+\exp\{-a_2\sinh(\ddot\xi a_3(a_4-a_5(\hat N-x)/\bar N)+a_3/2)\}]}, \quad (A1.12)$$

with $a_4$ and $a_5$ denoting constants, and $\gamma^+(x)$ follows from (A1.5) and (A1.6) as

$$\gamma^+(x) = cv(x)[1-\exp\{-m(x)(t-(x-1)/\bar N\hat n z)\}] + \varphi. \quad (A1.13)$$

The expression for $\gamma^-(x)$ will be given later.

The function $q(\xi,\gamma)$ given by (3.29) and (3.30) can be approximated reasonably accurately by the function

$$\bar q(\xi,\gamma) = 1-(1-q_o)\exp(-\rho_o(\xi)\gamma), \quad (A1.14)$$

where $\rho_o(\xi)$ is given by

$$\rho_o(\xi) = 53300 \frac{\check{\xi}^s - 1}{(\check{\xi}/\xi)^s - 1}, \quad 0.58 \leq \xi \leq 1.8. \tag{A1.15}$$

This approximation will enable us to derive a differential equation for $\psi$. Putting

$$\bar{c}x/\bar{N} = \tau, \tag{A1.16}$$

so that for $\bar{N}$ large we have $1/\bar{N} \approx 0$, and with r and $\hat{n}$ defined by

$$r \triangleq R/\bar{N}, \quad \hat{n} = \hat{N}/\bar{N}, \tag{A1.17}$$

we find that

$$\int_1^{N(t)} \exp\{\bar{c}x/\bar{N}\}dx \approx \bar{N}[\exp\{\bar{c}n\}-1]/\bar{c}, \tag{A1.18}$$

and

$$\int_1^{R(t)} \exp\{\bar{c}(\hat{N}+1-x)/\bar{N}\}dx \approx \bar{N}\exp\{\bar{c}\hat{n}\}[1-\exp\{-\bar{c}r\}]/\bar{c}. \tag{A1.19}$$

From the structure of (A1.6) we can infer that the function $\psi(\cdot)$ will be of the form

$$\psi(v,t) = cv(1-g(t)), \tag{A1.20}$$

where $g(t)$ has now to be found.

Using (A1.13)–(A1.20), Equations (A1.7) and (A1.8) become

$$[1-(1-q_o)\exp\{-\rho_o(\xi)cv(1-g(t))\}]h_n(\check{\xi})[\exp\{\bar{c}n\}-1] =$$

$$= \int_o^{\bar{c}n} h(\check{\xi},\tau)[\exp\{\tau\}-(1-q_o)\exp\{\tau-\rho_o(\xi)cv(1-\exp[-m(\tau)(t-\tau/\bar{c}\hat{n}z)])\} -$$

$$- \rho_o(\xi)\varphi(t)\}] d\tau, \tag{A1.21}$$

and

$$[1-(1-q_o)\exp\{-\rho_o(\xi)\varphi(t)\}]h_r(\check{\xi})[1-\exp\{-\bar{c}r\}] =$$

$$= \int_o^{\bar{c}r} h^*(\check{\xi},\tau)[\exp\{-\tau\}-(1-q_o)\exp\{-\tau-\rho_o(\xi)\gamma^-(\tau)\}]d\tau. \tag{A1.22}$$

We shall deal with (A1.21) first. Since for fixed values of $\mathring{\xi}$ the function $S(n)$ appearing in (A1.9) is independent of $q(\cdot)$ (and hence of $\psi(v,t)$) we must have that

$$h_n(\mathring{\xi}) \int_0^{\bar{c}n} \exp\{\tau\}d\tau = \int_0^{\bar{c}n} h(\mathring{\xi},\tau)\exp\{\tau\}d\tau, \qquad (A1.23)$$

and also

$$h_n(\mathring{\xi}) [\exp\{\bar{c}n\}-1]\exp\{-\rho_o(\mathring{\xi})cv(1-g(t))\} =$$

$$= \int_0^{\bar{c}n} h(\mathring{\xi},\tau)\exp\{\tau-\rho_o(\mathring{\xi})cv(1-\exp[-m(\tau)(t-\tau/\bar{c}\hat{n}z)])-\rho_o(\mathring{\xi})\varphi(t)\}d\tau. \qquad (A1.24)$$

Relation (A1.23) enables us to determine the function $S(n)$ in the following way. Since both functions, $\exp\{\tau\}$ and

$$h(\mathring{\xi},\tau) = \frac{1}{b_2[a_1+\exp\{-a_2\sinh(\mathring{\xi}a_3(a_4-a_5\tau/\bar{c})+a_3/2)\}]}, \qquad (A1.25)$$

are continuous in $[0,\bar{c}n]$ for all admissible values of $a_1, a_2, a_3, a_4, a_5, \mathring{\xi}$, and $\exp\{\tau\}$ is monotonically increasing, it follows from the generalized first mean value theorem that for all $n \in [0,1]$ there exists a $T(n) \in (0,n)$, such that (A1.23) holds, with $S(n)$ in $h_n(\mathring{\xi})$ given by

$$S(n) = a_4 - a_5 T(n). \qquad (A1.26)$$

The numerical quadrature of the right-hand side of (A1.23) for the domain of admissible parameter values $a_1,\ldots,\mathring{\xi}$ reveals that $S(n)$ is only very slightly dependent on $\mathring{\xi}$ for $\mathring{\xi} \in [-1.8, 0.6]$, which range is representative of normalized contraction velocities normally occurring in living muscle. It turns out that $S(n)$ can be closely approximated by

$$S(n) = a_4 - a_5 n(0.568+0.2307n) = \frac{a_2'}{B} - \frac{a_3'}{B} n(0.568+0.2307n), (A1.27)$$

where the constants $a_2'$, $a_3'$ and $B = 0.297$ have the same meaning as in Equation (4.29), i.e. they are related to the contraction time of the N-th motor unit.

Having established an expression for the function $S(n)$, we now turn our attention to (A1.24), which can be written as

$$h_n(\mathring{\xi}) [\exp\{\bar{c}n\}-1]\exp\{\rho_o(\mathring{\xi}) [cvg(t)+\varphi(t)]\} =$$

$$= \int_0^{\bar{c}n} h(\mathring{\xi},\tau)\exp\{\tau+\rho_o(\mathring{\xi})cv \exp[-m(\tau)(t-\tau/\bar{c}\hat{n}z)]\}d\tau. \qquad (A1.28)$$

Since the exponential function in the integrand of (A1.28) is continuous in $[0,\bar{c}n]$, we may apply the generalized first mean value theorem which states that there exist $T'(n) \in (0,n)$ such that the integral on the right-hand side of (A1.28) equals

$$h'(\dot{\xi}) \int_0^{\bar{c}n} \exp\{\tau + \rho_o(\xi)cv \exp[-m(\tau)(t-\tau/\bar{c}\hat{n}z)]\}d\tau,$$

where $h'(\dot{\xi})$ is given by (A1.25) with $\tau/\bar{c} = T'(n)$.

We shall now attempt to find a function $g(t)$ such that

$$a_4 - a_5 T'(n) \equiv S(n),$$

where $S(n)$ is defined by (A1.27). This implies that $h'(\dot{\xi}) \equiv h_n(\dot{\xi})$ and (A1.28) becomes

$$[\exp\{\bar{c}n\}-1]\exp\{\rho_o(\xi)[cvg(t)+\varphi(t)]\} =$$

$$= \int_0^{\bar{c}n} \exp\{\tau + \rho_o(\xi)cv \exp[-m(\tau)(t-\tau/\bar{c}\hat{n}z)]\}d\tau. \qquad (A1.29)$$

The function $\rho_o(\xi)$ is only a weak function of $\xi$ for $\xi \in [0.7, 1.3]$, which interval is the normal operating range of muscles in vivo. Hence during small time intervals we may regard $\rho_o(\xi) = \rho_o$ as approximately constant. Let it be assumed that $v(t)$ is a piecewise constant control on the same interval as the control $z(t)$ (it will be seen below that this assumption is not as restrictive as might appear). Let, therefore, $v$ and $z$ have the constant values $v_1$ and $z_1$ on the interval $[0,t_1)$, so that (A1.29) becomes

$$[\exp(\bar{c}n(t))-1]\exp\{\rho_o(cv_1 g(t)+\varphi(t))\} =$$

$$= \int_0^{\bar{c}n(t)} \exp\{\tau + \rho_o cv_1 \exp[-m(\tau)(t-\tau/\bar{c}\hat{n}z_1)]\}d\tau. \qquad (A1.30)$$

Generally, there is no hope of obtaining a non-differential analytical expression for $g(t)$ from (A1.30) which does not contain the integral, not even for constant $v$ and $z$. However, it is possible to derive a differential equation for $\psi$ which no longer contains the integral.

Differentiating (A1.30) with respect to $t$, and noting that $dn/dt = \hat{n}z_1$, we have

155

$$\bar{c}\hat{n}z_1\exp\{\bar{c}n+\rho_o(cv_1g+\varphi)\} + \rho_o(cv_1\dot{g}+\dot{\varphi})I = \bar{c}\hat{n}z_1\exp(\bar{c}n+\rho_ocv_1)-$$

$$-\rho_ocv_1\int_o^{\bar{c}n} m(\tau)\exp\{-m(\tau)(t-\tau/\bar{c}\hat{n}z_1) + \tau +\rho_ocv_1\exp[-m(\tau)(t-\tau/\bar{c}\hat{n}z_1)]\}d\tau\,,$$

(A1.31)

where I is the integral (A1.30). A second differentiation of (A1.30) with respect to $v_1$ yields

$$Ig = \int_o^{\bar{c}n} \exp\{-m(\tau)(t-\tau/\bar{c}\hat{n}z_1) + \tau +\rho_ocv_1\exp[-m(\tau)(t-\tau/\bar{c}\hat{n}z_1)]\}d\tau\,.$$

By putting (in the expression for I) $\bar{c}n = \tau$ on the left-hand and right-hand sides, and changing the dummy variable to y, the above expression can be written as

$$I(\tau)g = [\exp(\tau)-1]\exp\{\rho_o(cv_1g+\varphi)\}g = J(\tau) \quad (A1.32)$$

where $J(\tau)$ represents the integral on the right-hand side.

An integration by parts of the integral K appearing on the right-hand side of (A1.31) yields

$$K = [m(\tau)J(\tau)]_0^{\bar{c}n} - \int_o^{\bar{c}n} J(\tau)dm(\tau)$$

$$= g[m(\bar{c}n)\exp(\bar{c}n) - m(0) - \int_o^{\bar{c}n} \exp(\tau)dm(\tau)]\exp[\rho_o(cv_1g+\varphi)]$$

$$= g\exp[\rho_o(cv_1g+\varphi)]\int_o^{\bar{c}n} \exp(\tau)m(\tau)d\tau, \quad (A1.33)$$

since

$$\int_o^{\bar{c}n} \exp(\tau)dm(\tau) = m(\bar{c}n)\exp(\bar{c}n) - m(0) - \int_o^{\bar{c}n} \exp(\tau)m(\tau)d\tau,$$

and where (A1.32) has been used.

Applying again the generalized first mean value theorem we can find a function $m(n) = m(T^*(n))$, $T^* \in (0,n)$, such that

$$m(T^*(n))\int_o^{\bar{c}n} \exp(\tau)d\tau = \int_o^{\bar{c}n} \exp(\tau)m(\tau)d\tau. \quad (A1.34)$$

This is done in the following way. From Equation (4.30) we know that

$$m(\tau) = 1/(A_2 - A_3\tau/\bar{c}), \quad (A1.35)$$

which enables us to evaluate analytically the right-hand side of (A1.34). By putting $x = C - \tau$, where $C \triangleq \bar{c}A_2/A_3$, we obtain from (A1.34)

$$m(n) = \frac{C\exp(C)}{A_2[\exp(\bar{c}n)-1]} \int_{C-\bar{c}n}^{C} \exp(-x)dx/x =$$

$$= \frac{C\exp(C)}{A_2[\exp(\bar{c}n)-1]} [\int_{C-\bar{c}n}^{\infty} \exp(-x)dx/x - \int_{C}^{\infty} \exp(-x)dx/x], \quad (A1.36)$$

where we have decomposed the integral into the two exponential integrals $E_1(C-\bar{c}n)$ and $E_1(C)$ respectively. For $y \in [1,\infty)$, which includes the possible range of values of $(C-\bar{c}n)$ and $C$, the integral $E_1(y)$ is given (see Abramowitz and Stegun, 1968, p. 231) by

$$y\exp(y)E_1(y) = \mathcal{E}_1(y) + \varepsilon(y), \quad \varepsilon(y) < 5.10^{-5}, \quad (A1.37)$$

where

$$\mathcal{E}_1(y) = (y^2 + 2.334733y + 0.250621)/(y^2 + 3.330657y + 1.681534). \quad (A1.38)$$

Hence $y\exp(y)E_1(y) \approx \mathcal{E}_1(y)$, so that (A1.36) becomes

$$m(n) = \frac{1}{\exp(\bar{c}n)-1}[\frac{\exp(\bar{c}n)}{A_2-A_3n} \mathcal{E}_1(\bar{c}A_2/A_3-\bar{c}n) - \frac{\mathcal{E}_1(\bar{c}A_2/A_3)}{A_2}], \quad (A1.39)$$

which is the required analytical expression for the function $m(n)$.

Equation (A1.33) can thus be written as

$$K = g\exp[\rho_o(cv_1g+\varphi)]m(n)[\exp(\bar{c}n)-1]$$

which may now be substituted into (A1.31). By virtue of the relations $cv_1g = cv_1 - \psi$, and $cv_1\dot{g} = -\dot{\psi}$ (see (A1.20)), we find that

$$\dot{\psi} = m(n)(cv_1-\psi) +$$
$$+ \bar{c}\hat{n}z_1[1-\exp\{\rho_o(\psi-\varphi)\}]/\rho_o[1-\exp(-\bar{c}n)] + \dot{\varphi}, \quad (A1.40)$$

where $\varphi(t)$ is an absolutely continuous and piecewise differentiable function, as yet unspecified.

Equation (A1.40) is the required differential equation for $\psi$, subject to constant controls $v_1$, $z_1$, and with a constant value $\rho_o$ on the interval $[0,t_1)$. It can easily be shown that the solutions $\psi(t)$ and $\varphi(t)$, $t \in [0,\infty)$, of (A1.40) and (A1.58) below, are bounded such that

$$0 \leq \psi(t), \varphi(t) \leq c, \quad t \in [0,\infty). \quad (A1.41)$$

This implies that no finite escape times exist for these solutions; we are thus permitted to *define* the values of $\psi(t)$ and $\varphi(t)$, at the switching time $t_1$, as the continuation of the solutions for $t < t_1$ (Coddington and Levinson, 1955). Thus if at time $t_1$ the controls v and (or) z switch discontinously to new constant values $v_2$ and $z_2$, Equations (A1.40) and (A1.58) below can be integrated over $[t_1,t_2)$ using the new initial conditions $\psi(t_1)$, $\varphi(t_1)$, even though the right-hand sides of the respective differential equations do not exist at $t = t_1$. By continuing in this fashion the global solutions $\psi(t)$ and $\varphi(t)$ are obtained. For obvious reasons the index 1 on $z_1$ and $v_1$ may now be dropped and Equation (A1.40) will be stated in its final form once an expression for $\dot\varphi(t)$ has been derived.

Equation (A1.40) holds true strictly only for $z \geq 0$ because it contains the specific recruitment history as specified by (A1.6). It can, however, be shown that this influence of control history on present values of the state variables quickly fades out, so that the model (A1.40) remains valid also for $z < 0$, with z equated to zero in (A1.40), *provided* $z(t)$ *does not alternate in sign too rapidly*. This theoretical result is also borne out by simulation responses (Hatze, 1978a). It should, however, be kept in mind that the system is basically *hereditary* but that a proper consideration of this fact would lead to almost unmanageable complexity.

We turn now to the derivation of the equation for $\varphi(t)$. In analogy to (A1.6) we have

$$\gamma_j(t) = \psi(t)\exp[-m(t-(j-1)\triangle t(t))]. \quad (A1.42)$$

Applying to (A1.22) an argument similar to that which led from (A1.21) to (A1.23) we must have that

$$h_r(\dot\xi) \int_{\bar c n}^{\bar c(n+r)} \exp(\tau)d\tau = \int_{\bar c n}^{\bar c(n+r)} \exp(\tau)h^*(\dot\xi,\tau)d\tau, \quad (1.43)$$

where the original integration limits (see (A1.1)) have been used, $h_r(\dot\xi)$ is given by (A1.11), and $h^*(\dot\xi,\tau)$ by (A1.12) with $(\hat N-x)/\bar N$ replaced by $\tau/\bar c$. We now put $y = \tau - \bar c n$, so that (A1.43) becomes

$$h_r(\dot\xi) \int_0^{\bar c r} \exp(y)dy = \int_0^{\bar c r} \exp(y)h^*(\dot\xi,y)dy, \quad (A1.44)$$

where now

$$h^*(\dot\xi,y) = \frac{1}{b_2[a_1+\exp\{-a_2\sinh(\dot\xi\, a_3(a_4-a_5n-a_5y/\bar c)+a_3/2)\}]}. \quad (A1.45)$$

It can be seen that (A1.44) is exactly analogous to (A1.23) so that the solution function $S(n,r)$ appearing in $h_r(\dot{\xi})$ is immediately obtained from (A1.27) by subsituting $a_5r(0.568+0.2307r)$ for $a_5y/\bar{c}$ in (A1.45), i.e.

$$S(n,r) = a_4 - a_5(n+0.568r+0.2307r^2) = \frac{a_2'}{B} - \frac{a_3'}{B}(n+0.568r+0.2307r^2). \quad (A1.46)$$

On the other hand, similar to the relation existing between (A1.21) and (A1.29), Equation (A1.22) also implies that

$$[1-\exp(-\bar{c}r(t))]\exp(-\rho_o\varphi(t)) =$$

$$= \int_o^{\bar{c}r(t)} \exp\{-\tau-\rho_o\psi(t)\exp[-m(\tau)(t-\tau/\bar{c}\hat{n}(-z))]\}d\tau, \quad (A1.47)$$

where t is counted from the time when z became negative (switching off of motor units), and r(t) is obviously given by

$$\dot{r} = -\dot{n} = -\hat{n}z(t), \quad r(0) = 0, \quad 0 \leq r \leq 1. \quad (A1.48)$$

Equation (A1.47) arises from the fact that when a motor unit is switched off it is transferred from the N-population of stimulated units to the R-population of semi-active units and its initial concentration $\psi(t)$ begins to decline exponentially according to $\psi\exp(-mt)$. The same process applies to the next unit but with a time delay of $\Delta t$. Carrying on in this way and summing the corresponding active states, we obtain the integral expression (A1.47).

Differentiating (A1.47) with respect to time we obtain

$$\exp(-\rho_o\varphi)[-\rho_o\dot{\varphi}(1-\exp\{-\bar{c}r\})-\bar{c}\hat{n}z\exp(-\bar{c}r)] =$$

$$= \rho_o\psi \int_o^{\bar{c}r} m(\tau)\exp\{-m(\tau)(t-\tau/\bar{c}\hat{n}(-z)) -\tau-\rho_o\psi\exp[-m(\tau)(t-\tau/\bar{c}\hat{n}(-z))]\}d\tau -$$

$$- \bar{c}\hat{n}z\exp\{-\bar{c}r-\rho_o\psi\}-\rho_o\dot{\psi}J(t,\bar{c}r),$$

$$(A1.49)$$

where

$$J(t,\bar{c}r) = \int_o^{\bar{c}r} \exp\{-m(\tau)(t-\tau/\bar{c}\hat{n}(-z))-\tau-\rho_o\psi\exp[-m(\tau)(t-\tau/\bar{c}\hat{n}(-z))]\}d\tau. \quad (A1.50)$$

A second differentiation of (A1.47) with respect to $\rho_o$ yields

$$-\varphi[1-\exp(-\bar{c}r)]\exp(-\rho_o\varphi) = -\psi J(t,\bar{c}r). \quad (A1.51)$$

Integrating by parts the integral L on the right-hand side of (A1.49), we have

$$L = [m(n+r-\tau/\bar{c})J(t,\tau)]_0^{\bar{c}r} - \int_0^{\bar{c}r} J(t,\tau)dm(n+r-\tau/\bar{c})$$

$$= m(n)J(t,\bar{c}r) - \frac{\varphi}{\psi}\exp(-\rho_o\varphi)\int_0^{\bar{c}r}[1-\exp(-\tau)]dm(n+r-\tau/\bar{c})$$

$$= \frac{\varphi}{\psi}\exp(-\rho_o\varphi)\int_0^{\bar{c}r}\exp(-\tau)m(n+r-\tau/\bar{c})d\tau, \quad \text{(A1.52)}$$

where $m(n+r-\tau/\bar{c})$ means that m is a function of the argument $(n+r-\tau/\bar{c})$. Note that $n+r = \hat{n}$ is approximately constant. Again, by the generalized first mean value theorem there exists a $T(r) \in (0,r)$ such that

$$m(T(r),n)\int_0^{\bar{c}r}\exp(-\tau)d\tau = \int_0^{\bar{c}r}\exp(-\tau)m(n+r-\tau/\bar{c})d\tau. \quad \text{(A1.53)}$$

Put $y = \bar{c}(n+r) - \tau$. Then the right-hand side of (A1.53) becomes

$$\exp[-\bar{c}(n+r)]\int_{\bar{c}n}^{\bar{c}(n+r)}\exp(y)m(y/\bar{c})dy, \quad \text{(A1.54)}$$

where from (4.30)

$$m(y/\bar{c}) = 1/(A_2 - A_3 y/\bar{c}). \quad \text{(A1.55)}$$

Using the same procedure for (A1.53) as applied to (A1.34) we find that (for $m(n,r) \triangleq m(T(r),n)$)

$$m(n,r) = \frac{1}{1-\exp(-\bar{c}r)}\left[\frac{\mathcal{E}_1(\bar{c}A_2/A_3 - \bar{c}(n+r))}{A_2 - A_3(n+r)} - \frac{\exp(-\bar{c}r)}{A_2 - A_3 n}\mathcal{E}_1(\bar{c}A_2/A_3 - \bar{c}n)\right], \quad \text{(A1.56)}$$

where $\mathcal{E}_1(y)$ is given by (A1.38).

Hence (A1.52) becomes

$$L = \frac{\varphi}{\psi}\exp(-\rho_o\varphi)m(n,r)[1-\exp(-\bar{c}r)], \quad \text{(A1.57)}$$

which may be substituted into (A1.49) to yield

$$\dot{\varphi} = -m(n,r)\varphi + \dot{\psi}\varphi/\psi - \bar{c}\hat{n}z[1-\exp\{\rho_o(\varphi-\psi)\}]/\rho_o[\exp(\bar{c}r)-1], \quad \text{(A1.58)}$$

where $\dot\psi = m(n)[cv-\psi]$, for $z < 0$ (see (A1.40)).

When $z \geq 0$, no units are transferred from the N-population to the R-population and hence $\dot\varphi = -\varphi m(n,r)$, i.e. $\varphi$ declines exponentially.

From the preceding discussion on the domains of validity of Equations (A1.40) and (A1.58) it is clear that we need a *switching function* w(z) which switches parts of the equation on or off, depending on the value and sign of the control parameter z.

Let
$$w^+ = w(1+w)/2,$$
$$w^- = w(1-w)/2, \qquad (A1.59)$$

where the *switching function* $w = w(z)$ is expressed by
$$w \triangleq \text{sgn}(z), \qquad (A1.60)$$

i.e.
$$w = 1 \quad \text{for } z > 0,$$
$$w = 0 \quad \text{for } z = 0,$$
$$w = -1 \quad \text{for } z < 0.$$

Then it follows that (A1.40) and (A1.58) can respectively be written as

$$\dot\psi = m(n)(cv-\psi) + w^+ z\bar{c}\hat{n}[1-\exp\{\rho_o(\xi)(\psi-\varphi)\}]/\rho_o(\xi)[1-\exp(-\bar{c}n-\bar\delta)] -$$
$$- (1+w^-)m(n,r)\varphi, \quad \psi(0) = \psi_o, \qquad (A1.61)$$

and

$$\dot\varphi = -m(n,r)\varphi - w^-\{m(n)\varphi(cv/(\psi+\delta)-1)-$$
$$- z\bar{c}\hat{n}[1-\exp\{\rho_o(\xi)(\varphi-\psi)\}]/\rho_o(\xi)[\exp(\bar{c}r+\bar\delta)-1]\}, \quad \varphi(0) = \varphi_o, \qquad (A1.62)$$

since $\dot\psi\varphi/\psi = m(n)[cv-\psi]\varphi/\psi$ for $z < 0$. Note that the small constant $\delta$ (value about $10^{-8}$) has been added to the respective variables in order to obviate division by zero when $n = 0$, $r = 0$, $\psi = 0$. It can be shown that this procedure does not significantly influence the accuracy of the solution, provided $\delta$ is chosen sufficiently small. Computational difficulties which are due to the fact that $\psi > 0$ but $\varphi = 0$ in the last term of (A1.62) when $w^-$ becomes $-1$ for the first time, can be obviated by simply setting $\varphi(t_s) = \psi(t_s)$ at the switching time $t_s$ and then integrating in the normal way. After all, the difficulties arise only because $\varphi(t_s)$ should attain the value $\psi(t_s)$ in an infinitely short period of time.

Finally, we have to derive the differential equation defining r(t). Basically, this differential equation is given by (A1.48), but contradictions arise when during a contractive situation $\varphi > 0$, $r > 0$, but $z$ becomes zero and remains zero for a longer period of time. Then $\varphi(t)$ declines exponentially (by (A1.62)) but $r(t)$ remains constant (by (A1.48)) although, after some time, $\varphi$ will be practically zero and hence $r$ (the normalized population of semi-active motor units) should also be zero. This problem can be overcome by *defining* a unit to be *inactive* if $\varphi$ becomes smaller than a certain threshold value $k_2c$, i.e. we require that

$$r \to 0 \text{ as } \varphi \to k_2c. \quad (A1.63)$$

We now define the function $f(\varphi/k_2c)$ by

$$f(\varphi/k_2c) = (\varphi/k_2c)^2. \quad (A1.64)$$

For $z \geq 0$ we want $r(t)$ to decline rapidly when $\varphi(t) \leq k_2c$. Hence we augment (A1.48) by the term $(1+w^-)m'r$, i.e. (A1.48) becomes

$$\dot{r} = -\hat{n}z - (1 + w^-)m'r, \quad (A1.65)$$

where the last term is non-zero only for $z \geq 0$, and the rate function $m'$ is defined by the relation

$$m' = m(n,r)/f(\varphi/k_2c) = m(n,r)/[\Delta + (\varphi/k_2c)^2], \quad (A1.66)$$

where (A1.63)–(A1.64) have been used and a suitably small constant $\Delta$ has been added to provide for the case when $\varphi = 0$. It is not difficult to show that $\Delta = 10^{-3}m(n,r)$, and $k_2 = 10^{-4}$ constitute appropriate choices for the respective constants. In fact, with the above value for $k_2$, we have $k_2c = 1.373 \times 10^{-8}$ mole which is just the resting Ca-concentration in the muscle (Ebashi and Endo, 1968), i.e. a motor unit is declared inactive when its Ca-concentration has reached the actual physiological resting value.

A final point must be clarified. It is seen that upon integrating (A1.48) from $t = 0$ (where $z > 0$) to some $t$, the constraint on $r$ is violated, since $r$ will become negative. This can be prevented, without significantly affecting the accuracy of the solution, by simply multiplying $z(t)$ by $(r-w^-\bar{\delta})/(r+\bar{\delta})$, $\bar{\delta} = 10^{-8}$ (say). This procedure also removes the constraints on $r$.

With (A1.66) substituted into (A1.65) the final differential equation for $r$ thus becomes

$$\dot{r} = -\hat{n}z(r-w^-\bar{\delta})/(r+\bar{\delta}) - (1+w^-)m(n,r)r/[10^{-3}m(n,r) + (\varphi/k_2c)^2], \quad r(0) = r_o, \quad (A1.67)$$

which, together with (A1.4), constitutes the recruitment dynamics. Note also that the augmented Equation (A1.67) does not contradict

(A1.48), which was used in the derivation of (A1.58) for $z < 0$, since for this case (A1.67) reduces to (A1.48).

To complete the derivation of the model equations we return to Equation (A1.3a) which defines the function $h_o(\bar{\xi})$. Using an integral approximation to the sums we obtain

$$h_o(\bar{\xi}) \int_{\bar{c}(n+r)}^{\bar{c}} \exp(\tau)d\tau = \int_{\bar{c}(n+r)}^{\bar{c}} \exp(\tau)h(\bar{\xi},\tau)d\tau, \quad (A1.68)$$

where $h(\bar{\xi},\tau)$ is given by (A1.25).

Following the same procedure that led to (A1.27) we find that

$$S_o = \frac{a_2'}{B} - \frac{a_3'}{B}[n+r+0.568(1-n-r)+0.2307(1-n-r)^2], \quad (A1.69)$$

where $S_o$ is that function in the expression for $h_o(\bar{\xi})$ which corresponds to $S(n)$ in (A1.9).

We may again use integral approximations to the sums appearing in (4.36), and define

$$\varepsilon_n \triangleq q(\xi,\psi)[\exp(\bar{c}n)-1]/[\exp(\bar{c})-1], \quad (A1.70)$$

$$\varepsilon_r \triangleq q(\xi,\varphi)[\exp(\bar{c}n+\bar{c}r)-\exp(\bar{c}n)]/[\exp(\bar{c})-1], \quad (A1.71)$$

$$\varepsilon_o \triangleq q_o[\exp(\bar{c})-\exp(\bar{c}n+\bar{c}r)]/[\exp(\bar{c})-1], \quad (A1.72)$$

so that the *excitation function* $\varepsilon$ can be written as

$$\varepsilon = \varepsilon_n + \varepsilon_r + \varepsilon_o. \quad (A1.73)$$

Using (A1.2), (A1.3) and (A1.3a) together with the expressions (A1.70) – (A1.72) in (A1.1), it is seen that the latter can be represented as

$$F^C/\bar{F} = k(\bar{\xi})[\varepsilon_n h_n(\bar{\xi})+\varepsilon_r h_r(\bar{\xi})+\varepsilon_o h_o(\bar{\xi})]-b_1 k_1(\bar{\xi}). \quad (A1.74)$$

We now define the function $h(\bar{\xi})$ by

$$\varepsilon h(\bar{\xi}) \triangleq \varepsilon_n h_n(\bar{\xi})+\varepsilon_r h_r(\bar{\xi})+\varepsilon_o h_o(\bar{\xi}), \quad (A1.75)$$

where $\varepsilon$ is given by (A1.73), and the expression for $h(\bar{\xi})$ reads

$$h(\bar{\xi}) = \frac{1}{b_2[a_1+\exp\{-a_2\sinh(\bar{\xi}a_3 S+a_3/2)\}]}. \quad (A1.76)$$

The problem is to find a function $S$ which is such that (A1.75) is (at least approximately) satisfied. It can be shown that for any admissible value of $\bar{\xi}$, the maximally possible variations of $S(n)$, $S(n,r)$, and $S_o$ are sufficiently small to permit locally linear approximations of the functions $h_i(\dot{\xi})$, $i=n,r,o$ in (A1.75). By (A1.75) this implies that

$$S \approx (\varepsilon_n S(n) + \varepsilon_r S(n,r) + \varepsilon_o S_o)/\varepsilon, \qquad (A1.77)$$

which is the required expression for the function S appearing in (A1.76). The functions $\varepsilon_n, \varepsilon_r, \varepsilon_o, \varepsilon, S(n), S(n,r)$, and $S_o$ are defined by (A1.70) – (A1.73), (A1.27), (A1.46), and (A1.69) respectively.

Substituting (A1.75) into (A1.74), the latter becomes

$$F^C/\bar{F} = k(\xi)\varepsilon h(\dot{\xi}) - b_1 k_1(\xi), \qquad (A1.78)$$

which, if solved for $\dot{\xi}$, yields the last of Equations (4.37), since $F^C = F^{SE}$.

This completes the derivation of the model equations.

## APPENDIX A2

# Derivation of the model equations for penniform muscles

We shall use the notation introduced in Section 4.4 of the main text.
Let the muscle be at length $\bar{L}$ and relaxed. The the i-th fibre is inclined to the horizontal at an angle $\theta_{oi}$. We shall assume that the increment

$$\triangle \theta_{oi} \triangleq \theta_{o(i+1)} - \theta_{oi}$$

is uniform for all $\bar{\rho}$ fibres in the left half of the muscle, i.e.

$$\theta_{oi} = \check{\theta}_o + (\hat{\theta}_o - \check{\theta}_o)i/\bar{\rho}; \qquad (A2.1)$$

and that the ratio

$$k_{si} \triangleq \bar{\lambda}_i/\lambda_{soi}, \quad i=1,\ldots,\bar{\rho}, \qquad (A2.2)$$

of the optimum length $\bar{\lambda}_i$ of the $CE_i$ to the rest length $\lambda_{soi}$ of the tendinous part of the i-th fibre is approximately normally distributed with mean $\bar{k}_s$ and variance $\sigma_s^2$.

By virtue of (4.42) and (A2.2), condition (4.46) implies that for $L = \bar{L}$ and the muscle contracting under maximum stimulation we have that

$$[\exp\{\frac{\sigma}{\bar{\alpha}}(d/\lambda_{so}\cos\bar{\theta} - k_s - \bar{\kappa}k_s - 1)\} - 1]\sin\bar{\theta} =$$

$$= \frac{1}{\bar{\rho}} E\{\sum_{i=1}^{\bar{\rho}} [\exp\{\frac{\sigma}{\bar{\alpha}}(d/\lambda_{soi}\cos\bar{\theta}_i - k_{si} - \bar{\kappa}k_{si} - 1) - 1]\sin\bar{\theta}_i\} \qquad (A2.3)$$

$$= \frac{1}{\bar{\rho}} \sum_{i=1}^{\bar{\rho}} E\{\exp\{\frac{\sigma}{\bar{\alpha}}(d/\lambda_{soi}\cos\bar{\theta}_i - k_{si} - \bar{\kappa}k_{si} - 1)\}\}\sin\bar{\theta}_i - \frac{1}{\bar{\rho}} \sum_{i=1}^{\bar{\rho}} \sin\bar{\theta}_i,$$

where in analogy to (A2.2) $k_s = \bar{\lambda}/\lambda_{so}$, $E\{\cdot\}$ denotes the statistical expectation, and $\bar{\theta}$ and $\bar{\theta}_i$ denote respectively the inclination angles of the lumped model element and the i-th fibre under maximum isometric contraction at $L = \bar{L}$. Note that the step from the second to the third line in (A2.3) was possible bacause the $k_{si}$, $i=1,\ldots,\bar{\rho}$, are assumed uncorrelated.

Now
$$d/\lambda_{so}\cos\bar{\theta} = (\bar{\lambda}+\lambda_{s1})/\lambda_{so} = (\bar{\lambda}+(1+\bar{\alpha})\lambda_{so})/\lambda_{so} = k_s + 1 + \bar{\alpha},$$
and similarly
$$d/\lambda_{soi}\cos\bar{\theta}_i = k_{si} + 1 + \bar{\alpha},$$
which may be substituted into (A2.3).

Since $k_{si}$ and $\bar{\theta}_i$ are independent, (A2.3) implies that

$$\bar{\rho}\sin\bar{\theta} = \sum_{i=1}^{\bar{\rho}} \sin\bar{\theta}_i \tag{A2.4}$$

and

$$\exp(-\sigma\bar{\kappa}k_s/\bar{\alpha})\bar{\rho}\sin\bar{\theta} = \sum_{i=1}^{\bar{\rho}} E\{\exp(-\sigma\bar{\kappa}k_{si}/\bar{\alpha})\}\sin\bar{\theta}_i. \tag{A2.5}$$

Because the number of fibres $\bar{\rho}$ is usually very large, we may use an integral approximation to the sums, so that (A2.4) and (A2.5) become

$$\bar{\rho}\sin\bar{\theta} = \int_1^{\bar{\rho}} \sin\bar{\theta}(x)dx, \tag{A2.6}$$

and

$$\exp(-\sigma\bar{\kappa}k_s/\bar{\alpha})\bar{\rho}\sin\bar{\theta} = \int_1^{\bar{\rho}} E\{\exp(-\sigma\bar{\kappa}k_s(x)/\bar{\alpha})\}\sin\bar{\theta}(x)dx. \tag{A2.7}$$

Now
$$E\{\exp(-ak_s(x))\} =$$
$$= \int_{-\infty}^{\infty} \exp(-ak_s(x))(2\pi)^{-1/2}\sigma_s^{-1}\exp\{-\tfrac{1}{2}(\frac{k_s(x)-\bar{k}_s}{\sigma_s})^2\}dk_s(x), \tag{A2.8}$$

where $a = \sigma\bar{\kappa}/\bar{\alpha}$.

Combining the exponents in (A2.8) and completing the square, the right-hand side of (A2.8) can be written as

$$\exp\{-a\bar{k}_s + a^2\sigma_s^2/2\} \int_{-\infty}^{\infty} (2\pi)^{-1/2}\sigma_s^{-1} \exp\{-\frac{1}{2}(\frac{k_s(x)-(\bar{k}_s-a\sigma_s^2)}{\sigma_s})^2\} dk_s(x),$$

(A2.9)

where the integral is obviously equal to unity. Hence (A2.7) becomes

$$\exp(-ak_s)\bar{\rho} \sin \bar{\theta} = \exp(-a\bar{k}_s + a^2\sigma_s^2/2) \int_1^{\bar{\rho}} \sin \bar{\theta}(x) dx. \quad (A2.10)$$

By virtue of (A2.6) we see that

$$k_s = \bar{k}_s - a\sigma_s^2/2, \quad (A2.11)$$

where $a = \sigma\bar{\kappa}/\bar{\alpha}$. Relation (A2.11) means that the ratio $k_s = \bar{\lambda}/\lambda_{so}$ of the lumped model is equal to the mean ratio $\bar{k}_s$ of the individual fibres minus a term which depends on the variance $\sigma_s^2$ of the distribution.

We observe from Fig. 4.4 that if the muscle is held at length $L = \bar{L}$ and is at rest,

$$\tan \theta_{oi} = (\bar{\ell}_i - \lambda_{To})/d = \tan\{\check{\theta}_o + (\hat{\theta}_o - \check{\theta}_o)i/\bar{\rho}\}. \quad (A2.12)$$

If, on the other hand, $L = \bar{L}$ and the muscle is maximally stimulated, then

$$\tan \bar{\theta}_i = (\bar{\ell}_i - \lambda_{Ti})/d = (\bar{\ell}_i - \lambda_{To} - (\lambda_{Ti} - \lambda_{To}))/d = \tan \theta_{oi} - \bar{\alpha}\lambda_{To}/d. \quad (A2.13)$$

With (A2.12) substituted into (A2.13), Equation (A2.6) can be written as

$$\sin \bar{\theta} = (\hat{\theta}_o - \check{\theta}_o)^{-1} \int_{\tan \check{\theta}_o}^{\tan \hat{\theta}_o} (y + \bar{\alpha}\lambda_{To}/d) dy/(1+y^2)(y^2 + 2y\bar{\alpha}\lambda_{To}/d + (\bar{\alpha}\lambda_{To}/d)^2 + 1),$$

(A2.14)

where we have expressed $\sin \bar{\theta}$ in terms of $\tan \bar{\theta}$, and where

$$y \triangleq \tan\{\check{\theta}_o + (\hat{\theta}_o - \check{\theta}_o)x/\bar{\rho}\}.$$

The integral in (A2.14) can be solved analytically, but the resulting expression is exceedingly complicated. It can, however, be shown that the approximation

$$\bar{\theta} = \arctan\{\frac{1}{\bar{\rho}} \int_0^{\bar{\rho}} \tan\bar{\theta}(x) dx\} =$$

$$= \arctan\{\frac{1}{\bar{\rho}} \int_0^{\bar{\rho}} [\tan\{\check{\theta}_o+(\hat{\theta}_o-\check{\theta}_o)x/\bar{\rho}\}-\bar{\alpha}\lambda_{T_o}/d]\, dx\} \quad (A2.15)$$

introduces, for $\theta_o \in [0.52, 1.41]$, an average error of only 4%, and a maximum error of 17%. The range of $\theta_o$ corresponds to the range of inclinations normally occurring in penniform muscles.

The integral in (A2.15) is easily evaluated and yields

$$\bar{\theta} = \arctan\{(\ell n \cos \check{\theta}_o - \ell n \cos \hat{\theta}_o)/(\hat{\theta}_o-\check{\theta}_o)-\bar{\alpha}\lambda_{T_o}/d\}, \quad (A2.16)$$

where of course also

$$\theta_o = \arctan\{(\ell n \cos \check{\theta}_o - \ell n \cos \hat{\theta}_o)/(\hat{\theta}_o-\check{\theta}_o)\}. \quad (A2.17)$$

We have now succeeded in identifying the parameters $k_s$, $\bar{\theta}$, and $\theta_o$ of the lumped model in terms of the parameters $\bar{k}_s$, $\sigma_s^2$, $d$, $\check{\theta}_o$ and $\hat{\theta}_o$ of the distributed system. Strictly speaking, this approximation is valid only for $\theta = \bar{\theta}$, since (A2.3) is a special case of (4.46), corresponding to maximum isometric contraction at optimum length $\bar{L}$. However, $\theta$ is not constant during a general contractive mode, but changes according to

$$\theta(\ell,\lambda_T) = \arctan\{(\ell-\bar{\ell}-(\lambda_T-\lambda_{T_o}))/d + \tan \theta_o\}, \quad (A2.18)$$

as the lengths $\ell$ and $\lambda_T$, of the lumped model muscle and the tendon respectively, change. Hence the general form of (A2.3) should be resolved for all possible values of L, and all possible excitative states of all the muscle fibres. This is a fairly complicated approximation problem and would, if at all solvable, probably require a fundamental alteration in the structure of the lumped model. However, the approximation derived above is valid for $L = \bar{L}$, which is the most important operating point of the muscle. Practical experience has shown that with this model sufficiently accurate predictions are obtained over the whole range of possible values of $\ell$ and $\lambda_T$, and that hence there is no need for a more refined approximation.

A summary of the approximating expressions is given in Section 4.4 of the main text.

#### APPENDIX A3

# Computer program ELPEST for estimating myodynamic parameter values

The comments at the beginning of the listing of the program ELPEST fully explain its purpose and use. The program is written in ANSI FORTRAN IV and requires the external subroutine VAO5A which, in turn, requires the subroutine MB11. Since both routines are subject to copyright their listing cannot be supplied here. However, the user may substitute for VAO5A any routine which minimizes the sum of sqaures of p given functions, each of m variables. The calling statement of VAO5A must then be appropriately replaced by the calling statement of the substitute routine.

Below is a list of the names of the constants and variables that are used in the progran, and are also mentioned in the main text. Where text name and program name coincide, the definition will not be repeated here.

| | | |
|---|---|---|
| NDATA | = | p |
| ETO(ARRAY) | = | $\bar{N}(\phi_j, j=1,\ldots,p)$ |
| ANG(ARRAY) | = | $\phi_j (j=1,\ldots,p)$ |
| DIST(ARRAY) | = | $D_i(\phi_j, i=1,\ldots,m; j=1,\ldots,p)$ |
| EL(ARRAY) | = | $L_i(\phi_j,\ i=1,\ldots,m;\ j=1,\ldots,p)$, for penniform muscles (later in the program the real muscle lengths $L_i$ are replaced by the model muscle lengths $\ell_i$) |
| | = | $\ell_i\ (\phi_j,\ i=1,\ldots,m;\ j=1,\ldots,p)$, for fusiform muscles |
| ELBES(ARRAY) | = | $\bar{L}_i(i=1,\ldots,m)$, for penniform muscles (initial estimates of $\bar{L}_i$) |
| | | (later in the program the optimum real |

|  |  |  |
|---|---|---|
|  | = | muscle lengths $\bar{L}_i$ are replaced by optimum model muscle lengths $\bar{\ell}_i$) |
|  | = | $\bar{\ell}_i$ (i=1,...,m), for fusiform muscles (initial estimates of $\bar{\ell}_i$) |
| THEMIN(ARRAY) | = | $\check{\theta}_{oi}$(i=1,...,m), for penniform muscles |
|  | = | $\pi/2$, for fusiform muscles |
| THEMAO(ARRAY) | = | $\hat{\theta}_{oi}$(i=1,...,m), for penniform muscles |
|  | = | $\pi/2$, for fusiform muscles. |
| ALTO(ARRAY) | = | $\lambda_{Toi}$ (i=1,...,m) |
| ALB(ARRAY) | = | $\bar{\lambda}_i$ (i=1,...,m) |
| ALSO(ARRAY) | = | $\lambda_{soi}$(i=1,...,m) |
| THETAB(ARRAY) | = | $\bar{\theta}_i$(i=1,...,m) |
| SIGMA | = | $\sigma$ |

*Estimated parameter values:*

|  |  |  |
|---|---|---|
| PAR(J1) | = | $\bar{\alpha}_i$ (i=1,...,m) |
| PAR(J2) | = | $s_{ki}$ (i=1,...,m) |
| PAR(J3) | = | $\bar{F}_i$ (i=1,...,m) |
| PAR(J4) | = | $\bar{\ell}_i$ (i=1,...,m) |

For further explanations see the comments in the program listing.

```
      PROGRAM ELPEST
C***********************************************************************
C     THIS PROGRAM COMPUTES LEAST-SQUARES ESTIMATES OF MYODYNAMIC PARAMETER
C     VALUES RELATING TO THE PROPERTIES OF THE SERIES ELASTIC ELEMENTS AND
C     CONTRACTILE ELEMENTS OF A SET OF MUSCLES ACTING ACROSS A JOINT.
C
C     THE INPUT DATA ARE DERIVED FROM MAXIMUM ISOMETRIC TORQUE MEASUREMENTS
C     CARRIED OUT AT A SEQUENCE OF JOINT ANGLES. IT IS IMPORTANT THAT ALL
C     PASSIVE TORQUES DUE TO GRAVITY AND PASSIVE MUSCLE STRUCTURES HAVE BEEN
C     ZEROED OUT DURING THE MEASUREMENT, AND ARE THUS NOT INCLUDED IN THE
C     ISOMETRIC OUTPUT TORQUES.
C
C     INPUT PARAMETERS:
C
C     M    - NUMBER OF MUSCLES WHICH PRODUCE SIMULTANEOUSLY THE OBSERVED TORQUE
C            OUTPUT (MAXIMUM=10).
C     NDATA - NUMBER OF DISCRETE POINTS AT WHICH THE OUTPUT TORQUE HAS BEEN
C            OBSERVED (MAXIMUM=80). NOTE THAT NDATA MUST BE GREATER THAN OR
C            EQUAL TO 4*M, PREFERABLY ABOUT 15*M (CONSISTENT WITH THE LIMIT ON
C            NDATA).
C     IFUS - M-DIMENSIONAL VECTOR OF MUSCLE FORM INDICATOR. IFUS(I)=1 FOR
C            FUSIFORM MUSCLES, AND IFUS(I)=0 FOR MUSCLES WITH PENNIFORM FIBRE
C            ARRANGEMENT.
C     ETO  - NDATA-DIMENSIONAL VECTOR OF EXPERIMENTALLY OBSERVED VALUES OF THE
C            TOTAL MUSCULAR TORQUE OUTPUT (IN NEWTON*METER).
C     W    - NDATA-DIMENSIONAL VECTOR OF WEIGHTS FOR THE LEAST-SQUARE ESTIMATOR.
C            IN GENERAL, PUT W(I)=1. IF CERTAIN EXPERIMENTAL POINTS ARE OF SPECIAL
C            IMPORTANCE, ALLOCATE A VALUE OF W(I) GREATER THAN 1.0 TO THESE POINTS.
C            NOTE THAT THE WEIGHTS ARE DEFINED BY W(I)*(ETO(I)-MODELTORQUE(I)).
C     ANG  - NDATA-DIMENSIONAL VECTOR OF SUCCESSIVE (SMALLEST TO LARGEST) VALUES
C            OF JOINT ANGLES (IN RADIANS) AT WHICH THE TORQUE OBSERVATIONS WERE
C            MADE. THE VECTOR ANG(I), I=1,NDATA, MAY CONTAIN ONLY ZEROS, IN WHICH
C            CASE THE (NON-ZERO) COMPONENTS OF THE VECTORS DIST AND EL (SEE
C            BELOW) ARE READ IN. IF, HOWEVER, THE ARRAY ANG CONTAINS NON-ZERO
C            ELEMENTS, THE PROGRAM ASSUMES THAT THE VALUES OF THE ARRAY ELEMENTS
C            OF DIST AND EL ARE COMPUTED BY THE USER-SUPPLIED SUBROUTINES DICOMP
C            AND ELCOMP RESPECTIVELY. IF EQUIDISTANT MEASUREMENTS ARE MADE,
C            THEN ONLY THE FIRST ELEMENT OF ARRAY ANG NEED BE SPECIFIED, AND THE
C            REMAINING ELEMENTS MAY BE EQUATED TO ZERO. IN THIS CASE THE
C            SUBROUTINE ANCOMP AUTOMATICALLY COMPUTES THE REMAINING VALUES, AND
C            THE INCREMENT, DELANG, OF THE ANGULAR POSITIONS MUST BE READ IN BY
C            THE USER. NOTE, HOWEVER, THAT THE ABSOLUTE VALUE OF THE FIRST
C            ELEMENT OF ARRAY ANG MUST BE GREATER THAN 1.E-9. IF ANG(1)=0
C            ACCORDING TO THE EXPERIMENTAL RECORD, PUT ANG(1)=0.001 WHICH INTRO-
C            DUCES A NEGLIGIBLE ERROR.
C     DIST - (M,NDATA)-DIMENSIONAL MATRIX OF MEASURED MUSCULAR MOMENT ARM VALUES
C            CORRESPONDING TO THE JOINT ANGLES AT WHICH THE MEASUREMENTS WERE
C            MADE (MOMENT ARM VALUES IN METER). IF A FORMULA FOR THE COMPUTATION
C            OF THE ELEMENTS OF DIST IS AVAILABLE, SUBROUTINE DICOMP IS USED TO
C            FIND THESE VALUES. IN THIS CASE, NO VALUES HAVE TO BE READ IN FOR
C            DIST, BUT THE ARRAY ANG (SEE ABOVE) MUST BE GIVEN.
C     EL   - (M,NDATA)-DIMENSIONAL MATRIX OF MUSCLE LENGTHS CORRESPONDING TO
C            THE MEASURING JOINT ANGLES (LENGTHS IN METER). FOR PENNIFORM MUSCLES
C            EL IS THE DISTANCE FROM THE ORIGIN OF THE MOST PROXIMALLY LOCATED
C            FIBRE TO THE INSERTION OF THE COMMON TENDON AT THE DISTAL END. THE
C            LENGTH OF THE LUMPED MODEL MUSCLE IS THEN COMPUTED FROM EL(LUMPED
C            MODEL)=EL(REAL MUSCLE) - D*(TAN(THETAMAXO)-TAN(THETAO)). NOTE THAT
C            FOR PENNIFORM MUSCLES D=(ELBAR-LAMDATO)/TAN(THETAMAXO). SEE ALSO NOTE
C            ON ELBES BELOW. IF A COMPUTATIONAL FORMULA FOR EL(I,J) IS AVAILABLE,
C            IT MUST BE CODED IN SUBROUTINE ELCOMP. IN THIS CASE, NO VALUES HAVE
C            TO BE READ IN FOR EL.
C     FBEST - M-DIMENSIONAL VECTOR OF INITIAL ESTIMATES OF THE MAXIMUM ISOMETRIC
C            FORCES FBAR(I), I=1,...,M (NEWTONS). THIS ESTIMATE SHOULD
C            CORRESPOND TO THE VALUE EXPECTED FROM THE MAXIMUM PHYSIOLOGICAL
C            CROSS-SECTIONAL AREA OF THIS MUSCLE (TAKE APPROX. 100N/CM**2).
C            NOTE THAT THE PROGRAM CONSTRAINS THE COMPUTED ESTIMATE OF FBAR(I)=
C            PAR(J3), I=1,...,M, TO BE LESS THAN OR EQUAL TO FBEST(I). IF THE
C            USER DESIRES TO REMOVE THIS CONSTRAINT, HE SHOULD SIMPLY INCREASE
C            THE VALUE OF FBEST(I) TO TWICE OR THREE TIMES THE ORIGINAL VALUE.
C     ELBES - M-DIMENSIONAL VECTOR OF INITIAL ESTIMATES OF THE OPTIMAL MUSCLE
C            LENGTHS LBAR(I), I=1,...,M (METER).
C     THEMIN - M-DIMENSIONAL VECTOR OF THE VALUES OF THE MINIMUM MUSCLE FIBRE
```

```
C                    INCLINATION (RADIANS). FOR FUSIFORM MUSCLES THEMIN(I)=1.570796,
C                    WHILE FOR PENNIFORM MUSCLES THIS VALUE DEPENDS ON THE FIBRE
C                    ARRANGEMENT AND MUST BE ESTIMATED FROM THE MUSCLE GEOMETRY, WITH
C                    THE MUSCLE AT LENGTH ELBAR AND AT REST.
C           THEMAO - M-DIMENSIONAL VECTOR OF THE VALUES OF THE MAXIMUM MUSCLE FIBRE
C                    INCLINATION (RADIANS). FOR FUSIFORM MUSCLES THEMAO(I)=1.570796,
C                    WHILE FOR PENNIFORM MUSCLES THIS VALUE DEPENDS ON THE FIBRE
C                    ARRANGEMENT AND MUST BE ESTIMATED FROM THE MUSCLE GEOMETRY, WITH
C                    THE MUSCLE AT LENGTH ELBAR (APPROXIMATELY) AND AT REST.
C           ALTO   - M-DIMENSIONAL VECTOR OF REST LENGTHS OF THE TENDINOUS SERIES
C                    ELASTIC PARTS OF PENNIFORM AND FUSIFORM MUSCLES. FOR FUSIFORM
C                    MUSCLES, ALTO DENOTES THE REST LENGTHS OF THE TENDINOUS PARTS OF
C                    THE FIBRES, WHILE FOR PENNIFORM MUSCLES IT DENOTES THE REST LENGTH
C                    OF THE COMMON TENDON, NOT INCLUDING THE TENDINOUS PARTS OF THE
C                    FIBRES. THE LATTER ARE ASSUMED TO HAVE A FIXED LENGTH OF
C                    0.07*LAMDABAR FOR ALL PENNIFORM MUSCLES.
C           LASOIN - M-DIMENSIONAL VECTOR OF INTEGER VALUES (0 OR 1) FOR PENNIFORM
C                    MUSCLES. IF THE INDIVIDUAL FIBRES ARE ASSUMED TO CONTAIN NO
C                    TENDINOUS PARTS, AND ALL TENDINOUS MATERIAL RESIDES IN THE COMMON
C                    TENDON, PUT LASOIN(I)=0. OTHERWISE PUT LASOIN(I)=1, EXCEPT FOR
C                    VERY LONG TENDINOUS FIBRE PARTS, ALWAYS PUT LASOIN(I)=0.
C           MAXIT  - INTEGER VARIABLE INDICATING MAXIMUM NUMBER OF ITERATIONS PERMITTED
C                    IN THE LEAST-SQUARE MINIMIZATION ROUTINE. A VALUE OF 500 IS
C                    USUALLY SUFFICIENT.
C           IPR    - INTEGER VARIABLE. IF IPR=0, NO PRINT-OUT OF INTERMEDIATE RESULTS
C                    FROM THE OPTIMIZATION ROUTINE IS OBTAINED. IN THIS CASE, IPR IS USED
C                    AS AN OUTPUT PARAMETER (IPR=1 MEANS THAT THE SUM OF SQUARES FAILS TO
C                    DECREASE, IPR=2 MEANS THAT MAXIT ITERATIONS HAVE BEEN MADE). IF
C                    IPR IS SET EQUAL TO 1, THE VALUES OF VARIABLES AND FUNCTIONS ARE
C                    PRINTED OUT AT EACH ITERATION. IF IPR=-K, K A POSITIVE INTEGER,
C                    THE VALUES OF THE VARIABLES ARE PRINTED EVERY K ITERATION.
C           ACC    - REAL VARIABLE SPECIFYING REQUIRED ACCURACY IN THE CALCULATED SUM
C                    OF SQUARES. PUT ACC=0.0002 (APPROXIMATELY).
C           HSTEP  - REAL VARIABLE USED AS ARGUMENT INCREMENT IN THE ESTIMATION OF
C                    PARTIAL DERIVATIVES. THE VALUE OF HSTEP SHOULD BE BETWEEN 0.001
C                    AND 0.02 (PREFERABLY ABOUT 0.008). CONVERGENCE OF THE OPTIMIZATION
C                    ALGORITHM MAY DEPEND CRITICALLY ON THIS VALUE.
C
C        ESTIMATED PARAMETER VALUES:
C
C           PAR(J1) - VALUES OF THE CONSTANTS ALPHABAR(I), I=1,...,M.
C           PAR(J2) - VALUES OF THE CONSTANTS STAN(I), I=1,...,M. THESE VALUES ARE THE
C                     STANDARD DEVIATIONS OF THE LENGTH-TENSION CURVES OF THE MUSCLES.
C           PAR(J3) - VALUES OF THE CONSTANTS FBAR(I), I=1,...,M.
C           PAR(J4) - VALUES OF THE CONSTANTS LBAR(I), I=1,...,M. FOR PENNIFORM
C                     MUSCLES, THESE VALUES CORRESPOND TO THE OPTIMUM LENGTHS OF THE
C                     LUMPED MODEL MUSCLES.
C
C        EXTERNAL SUBROUTINES USED - VA05A,MB11. ALL ROUTINES IN HATLIB.
C
C        USER-SUPPLIED SUBROUTINES -
C        FOR AN EXPLANATION OF THE PURPOSE OF THESE ROUTINES SEE ANG, DIST, AND EL
C        ABOVE.
C           DICOMP(M,NDATA) - ROUTINE REQUIRES COMMON/EXFUV/.
C           ELCOMP(M,NDATA) - ROUTINE REQUIRES COMMON/EXFUV/.
C
C************************************************************************
      DIMENSION F2(80),WKS2(9960),PMI(40)
C     THE FOLLOWING COMMON CONTAINS THE EXPERIMENTALLY OBSERVED TORQUE OUTPUT
C     VALUES ETO, THE MOMENT ARM MATRIX DIST, AND THE MUSCLE LENGTHS MATRIX EL.
      COMMON/EXFUV/ETO(80),DIST(10,80),EL(10,80),ANG(80)
      COMMON/FOCOS/I1,K,M,NPAR,TAD(10),EXSIG(10),SAL(10),ALB(10),
     *             ALSO(10),IFUS(10),THETAB(10),D(10),ALTO(10),ELB(10),
     *             FBEST(10),ELBES(10),THEMAO(10),TATHEO(10),THEMIN(10),
     *             SIGMA,LASOIN(10),SMIN
      COMMON/ESTPAR/PAR(40)
      COMMON/TLREL/TLX(10),PENFAC(10)
      COMMON/WRFILE/LR,LW
      COMMON/TORQUE/TOTOR(80),ICONST,W(80)
      EXTERNAL FCT
```

```fortran
C
      LR=5
      LW=6
      READ(LR,10) M,NDATA,DELANG
      READ(LR,20) (IFUS(I),I=1,M)
      READ(LR,30) (ETO(I),I=1,NDATA)
      READ(LR,25) (W(I),I=1,NDATA)
      READ(LR,35) (ANG(I),I=1,NDATA)
C
      ANGTOT=0.
      DO 11 J=1,NDATA
   11 ANGTOT=ANGTOT+ABS(ANG(J))
      IF(ANGTOT.LT.1.E-10) GO TO 12
      IF((ABS(ANG(2))+ABS(ANG(3))).LT.1.E-10) CALL ANCOMP(DELANG,NDATA)
      CALL DICOMP(M,NDATA)
      CALL ELCOMP(M,NDATA)
      GO TO 13
C
   12 READ(LR,40) ((DIST(I,J),I=1,10),J=1,NDATA)
      READ(LR,40) ((EL(I,J),I=1,10),J=1,NDATA)
   13 READ(LR,50) (FBEST(I),I=1,M)
      READ(LR,40) (ELBES(I),I=1,M)
      READ(LR,35) (THEMIN(I),I=1,M)
      READ(LR,35) (THEMAO(I),I=1,M)
      READ(LR,40) (ALTO(I),I=1,M)
      READ(LR,20) (LASOIN(I),I=1,M)
      READ(LR,45) MAXIT,IPR,ACC,HSTEP
      WRITE(LW,80)
      WRITE(LW,85)
      WRITE(LW,90) M,NDATA,DELANG
      WRITE(LW,95)
      WRITE(LW,100) (IFUS(I),I=1,M)
      WRITE(LW,105)
      WRITE(LW,110) (ETO(I),I=1,NDATA)
      WRITE(LW,280)
      WRITE(LW,140) (W(I),I=1,NDATA)
      WRITE(LW,270)
      WRITE(LW,140) (ANG(I),I=1,NDATA)
      WRITE(LW,170)
      WRITE(LW,120) ((DIST(I,J),I=1,10),J=1,NDATA)
      WRITE(LW,180)
      WRITE(LW,120) ((EL(I,J),I=1,10),J=1,NDATA)
      WRITE(LW,190)
      WRITE(LW,130) (FBEST(I),I=1,M)
      WRITE(LW,200)
      WRITE(LW,140) (ELBES(I),I=1,M)
      WRITE(LW,210)
      WRITE(LW,140) (THEMIN(I),I=1,M)
      WRITE(LW,220)
      WRITE(LW,140) (THEMAO(I),I=1,M)
      WRITE(LW,230)
      WRITE(LW,140) (ALTO(I),I=1,M)
      WRITE(LW,240)
      WRITE(LW,100) (LASOIN(I),I=1,M)
      WRITE(LW,250)
      WRITE(LW,260) MAXIT,IPR,ACC,HSTEP
      NPAR=4
      N=NPAR*M
      SIGMA=1.531
      SMIN=0.28
      ICONST=1
C
C     INITIAL ESTIMATES OF THE PARAMETERS, AND INITIAL COMPUTATIONS
C
      DO 1 I1=1,M
      IF(IFUS(I1).EQ.1) GO TO 9
      TATHEO(I1)=(ALOG(COS(THEMIN(I1)))-ALOG(COS(THEMAO(I1))))/(THEMAO(
     *           I1)-THEMIN(I1))
C     OBTAIN INITIAL ESTIMATE OF D AND OF VARIABLES DEPENDING ON D.
      D(I1)=(ELBES(I1)-ALTO(I1))/TAN(THEMAO(I1))
      TAD(I1)=D(I1)*(TAN(THEMAO(I1))-TATHEO(I1))
C     THE FOLLOWING STATEMENT REPLACES, FOR PENNIFORM MUSCLES, THE INITIAL
```

```
C     ESTIMATE ELBES(I1) OF THE OPTIMUM LENGTH LBAR(I1) OF THE LONGEST FIBRE BY
C     THE OPTIMUM LENGTH OF THE LUMPED MODEL MUSCLE.
      ELBES(I1)=ELBES(I1)-TAD(I1)
C     REPLACE REAL MUSCLE LENGTHS BY LUMPED MODEL MUSCLE LENGTHS.
      DO 2 K=1,NDATA
    2 EL(I1,K)=EL(I1,K)-TAD(I1)
C     INTRODUCE ELB(I1) (=LBAR(I1)) FOR THE LUMPED MODEL MUSCLE IN ORDER TO KEEP
C     ELBES(I1) CONSTANT. ELB(I1) WILL BE CHANGED BY THE PROGRAM IN THE PROCESS
C     OF ESTIMATING THIS PARAMETER.
      ELB(I1)=ELBES(I1)
C
C     INITIAL ESTIMATES OF PARAMETERS
C     NOTE THAT STAN(I) IS CONSTRAINED TO BE GREATER THAN OR EQUAL TO SMIN,
C     WHICH IS THE SMALLEST VALUE FOUND IN HUMAN MUSCLE. TO REMOVE THIS
C     CONSTRAINT WE DEFINE A NEW PAR(J2) BY THE RELATION STAN(I)=SMIN*(1.+
C     EXP(PAR(J2)**3)). A GOOD INITIAL ESTIMATE FOR THIS NEW PAR(J2) IS -1.
C     SIMILARLY, FBAR(I) IS CONSTRAINED BY THE MAXIMUM PHYSIOLOGICAL CROSS-SEC-
C     TIONAL AREA OF THE MUSCLE, TO WHICH THE ESTIMATED VALUE OF FBEST(I)
C     CORRESPONDS. TO REMOVE THIS CONSTRAINT WE DEFINE A NEW PAR(J3) BY
C     FBAR(I)=FBEST(I)/(1.+EXP(-PAR(J3))). A GOOD INITIAL ESTIMATE FOR THIS NEW
C     PAR(J3) IS 1. IF, HOWEVER, THE PARAMETER ICONST=0, THEN THIS CONSTRAINT
C     ON FBAR(I) IS NOT ACTIVE.
C
    9 CONTINUE
      J1=NPAR*(I1-1)+1
      J2=J1+1
      J3=J2+1
      J4=J3+1
      PAR(J1)=0.07
      PAR(J2)=-1.
      PAR(J3)=FBEST(I1)
      IF(ICONST.NE.0) PAR(J3)=1.
      PAR(J4)=ELB(I1)
      EXSIG(I1)=EXP(SIGMA)-1.
      IF(IFUS(I1).EQ.1) GO TO 6
      THETAB(I1)=ATAN(TATHEO(I1)-PAR(J1)*ALTO(I1)/D(I1))
      ALB(I1)=D(I1)/(COS(THETAB(I1))*(0.07*PAR(J1)+1.07))
      IF(LASOIN(I1).EQ.0) ALB(I1)=D(I1)/COS(THETAB(I1))
      ALSO(I1)=0.07*ALB(I1)
      GO TO 8
    6 ALSO(I1)=0.98*ALTO(I1)+0.02*PAR(J4)
      ELB(I1)=ELBES(I1)
      ALB(I1)=(ELB(I1)-(1.+PAR(J1))*ALSO(I1))
    8 SAL(I1)=SIGMA/(PAR(J1)*ALSO(I1))
    1 CONTINUE
C     NORMALIZE PARAMETER VALUES FOR USE IN MINIMIZATION ROUTINE. THE ROUTINE
C     REQUIRES THAT ALL PARAMETER VALUES HAVE SIMILAR MAGNITUDE. THIS IS
C     ACHIEVED HERE BY MAKING THEIR AVERAGE VALUES EQUAL TO 1. (-1. FOR PAR(J2))
      DO 4 I=1,M
      J1=NPAR*(I-1)+1
      J2=J1+1
      J3=J2+1
      J4=J3+1
      PMI(J1)=PAR(J1)/0.07
      PMI(J2)=PAR(J2)
      PMI(J3)=PAR(J3)/FBEST(I)
      IF(ICONST.NE.0) PMI(J3)=PAR(J3)
    4 PMI(J4)=PAR(J4)/ELBES(I)
      DMAX=0.20*SQRT(FLOAT(N))
      CALL VA05A(FCT,NDATA,N,F2,PMI,HSTEP,DMAX,ACC,MAXIT,IPR,WKS2)
C     DENORMALIZE ESTIMATED PARAMETER VALUES
      DO 5 I=1,M
      IF(LASOIN(I).EQ.0) ALSO(I)=0.
      J1=NPAR*(I-1)+1
      J2=J1+1
      J3=J2+1
      J4=J3+1
      PAR(J1)=PMI(J1)*0.07
      PAR(J2)=SMIN*(1.+EXP(PMI(J2)**3))
      PAR(J3)=PMI(J3)*FBEST(I)
      IF(ICONST.NE.0) PAR(J3)=FBEST(I)/(1.+EXP(-PMI(J3)))
    5 PAR(J4)=PMI(J4)*ELBES(I)
```

```
C
          WRITE(LW,150)
          WRITE(LW,155)
          I2=1
          DO 7 J=1,M
          I3=J*4
          WRITE(LW,160)J,ALSO(J),SIGMA,(PAR(I),I=I2,I3),ALB(J),D(J),THETAB(J
         *)
        7 I2=I2+4
   10     FORMAT(2I2,F10.8)
   20     FORMAT(10I1)
   25     FORMAT(16F5.1)
   30     FORMAT(8F10.2)
   35     FORMAT(16F5.3)
   40     FORMAT(10F5.4)
   45     FORMAT(I3,I3,4X,F10.8,F10.8)
   50     FORMAT(10F6.0)
   80     FORMAT(1H1,5X,11H INPUT DATA,//)
   85     FORMAT(3X,2H M,2X,6H NDATA,2X,7H DELANG)
   90     FORMAT(3X,I2,3X,I2,3X,F10.8,/)
   95     FORMAT(3X,17H IFUS(I),I=1,...,M)
  105     FORMAT(/3X,51H MEASURED TOTAL MUSCULAR TORQUE ETO(J),J=1,...,NDATA)
  100     FORMAT((3X,10(I1,4X)),/)
  110     FORMAT((3X,10(F6.2,3X)),/)
  120     FORMAT((3X,10(F5.4,4X)),/)
  130     FORMAT((3X,10(F5.0,4X)),/)
  140     FORMAT((3X,10(F5.3,4X)),/)
  150     FORMAT(//5X,27H ESTIMATED PARAMETER VALUES,//)
  155     FORMAT(1X,13HMUSCLE-NUMBER,3X,7HLAMDASO,5X,5HSIGMA,2X,8HALPHABAR,6
         1X,4HSTAN,6X,4HFBAR,6X,4HLBAR,2X,8HLAMDABAR,6X,1HD,5X,8HTHETABAR)
  160     FORMAT(6X,I2,8X,4(F8.3,2X),2X,F7.1,4(F8.3,2X))
  170     FORMAT(/3X,64HMEASURED MUSCLE MOMENT ARMS DIST(I,J), I=1,...,10, J
         1=1,...,NDATA)
  180     FORMAT(/3X,58HMEASURED MUSCLE LENGTHS EL(I,J), I=1,...,10, J=1,...
         1,NDATA)
  190     FORMAT(/3X,53HESTIMATED MAXIMUM ISOMETRIC FORCES FBEST(I),I=1,...,
         1M)
  200     FORMAT(/3X,51HESTIMATED OPTIMUM MUSCLE LENGTHS ELBES(I) I=1,...,M)
  210     FORMAT(/3X,22HTHETAMIN(I), I=1,...,M)
  220     FORMAT(/3X,23HTHETAMAXO(I), I=1,...,M)
  230     FORMAT(/3X,21HLAMDATO(I), I=1,...,M)
  240     FORMAT(/3X,20HLASOIN(I), I=1,...,M)
  250     FORMAT(/3X,41HMAXIT      IPR         ACC          HSTEP)
  260     FORMAT(/4X,I3,6X,I3,6X,F10.8,4X,F10.8)
  270     FORMAT(/3X,21HANG(I), I=1,...,NDATA)
  280     FORMAT(/3X,21HLEAST-SQUARES WEIGHTS)
          STOP
          END
          SUBROUTINE FCT(NDATA,N,F2,PMI)
C*****************************************************************
C         THIS ROUTINE COMPUTES THE DIFFERENCE BETWEEN THE OUTPUT TORQUE VALUES AS
C         PREDICTED BY THE MODEL, AND THE MEASURED TORQUE VALUES, AT EACH SAMPLING
C         POINT OF THE EXPERIMENTAL DATA.
C*****************************************************************
          DIMENSION PMI(40),F2(NDATA),XI(10),XAM(10)
          COMMON/EXFUV/ETO(80),DIST(10,80),EL(10,80),ANG(80)
          COMMON/FOCOS/I1,K,M,NPAR,TAD(10),EXSIG(10),SAL(10),ALB(10),
         *          ALSO(10),IFUS(10),THETAB(10),D(10),ALTO(10),ELB(10),
         *          FBEST(10),ELBES(10),THEMAO(10),TATHEO(10),THEMIN(10),
         *          SIGMA,LASOIN(10),SMIN
          COMMON/ESTPAR/PAR(40)
          COMMON/TLREL/TLX(10),PENFAC(10)
          COMMON/WRFILE/LR,LW
          COMMON/TORQUE/TOTOR(80),ICONST,W(80)
          EXTERNAL FCXIX,FCFIB
          MAXIT1=200
          MAXIT2=200
C         DENORMALIZE PARAMETER VALUES FOR COMPUTATIONS IN THIS ROUTINE AND
C         RECOMPUTE CONSTANTS FOR IMPROVED PARAMETER VALUES. PAR(J2) NOW BECOMES THE
C         ACTUAL STANDARD DEVIATION STAN(I).
          DO 5 I1=1,M
```

```
              J1=NPAR*(I1-1)+1
              J2=J1+1
              J3=J2+1
              J4=J3+1
              PAR(J1)=PMI(J1)*0.07
              PAR(J2)=SMIN*(1.+EXP(PMI(J2)**3))
              PAR(J3)=PMI(J3)*FBEST(I1)
              IF(ICONST.NE.0) PAR(J3)=FHEST(I1)/(1.+EXP(-PMI(J3)))
              PAR(J4)=PMI(J4)*ELBES(I1)
              IF(IFUS(I1).EQ.1) GO TO 7
              ELB(I1)=PAR(J4)
              D(I1)=(ELB(I1)-ALTO(I1))/TATHEO(I1)
       C      RECONSTRUCT REAL MUSCLE LENGTHS
              DO 9 KK=1,NDATA
            9 EL(I1,KK)=EL(I1,KK)+TAD(I1)
       C      COMPUTE NEW TAD(I1)
              TAD(I1)=D(I1)*(TAN(THEMAO(I1))-TATHEO(I1))
       C      REPLACE REAL MUSCLE LENGTHS BY LUMPED MODEL MUSCLE LENGTHS FOR NEW TAD(I1)
              DO 10 KK=1,NDATA
           10 EL(I1,KK)=EL(I1,KK)-TAD(I1)
              THETAB(I1)=ATAN(TATHEO(I1)-PAR(J1)*ALTO(I1)/D(I1))
              ALB(I1)=D(I1)/(COS(THETAB(I1))*(0.07*PAR(J1)+1.07))
              IF(LASOIN(I1).EQ.0) ALB(I1)=D(I1)/COS(THETAB(I1))
              ALSO(I1)=0.07*ALB(I1)
              GO TO 8
            7 ALSO(I1)=0.98*ALTO(I1)+0.02*PAR(J4)
              ELB(I1)=PAR(J4)
              ALB(I1)=(ELB(I1)-(1.+PAR(J1))*ALSO(I1))
            8 SAL(I1)=SIGMA/(PAR(J1)*ALSO(I1))
            5 CONTINUE
       C
              DO 1 K=1,NDATA
              TOTOR(K)=0.
              DO 2 I1=1,M
              J1=NPAR*(I1-1)+1
              J2=J1+1
              J3=J2+1
              J4=J3+1
              IF(IFUS(I1).NE.1) GO TO 3
       C      COMPUTATIONS FOR FUSIFORM MUSCLES
       C      COMPUTE XI(I1) FOR ALL FUSIFORM MUSCLES, FOR THE CURRENT VALUE OF K.
              CALL FZERO(FCXIX,0.2,2.0,F1,XI(I1),MAXIT1,MES1)
              IF(MES1.EQ.1) WRITE(LW,20)
              GO TO 4
       C      COMPUTATIONS FOR PENNIFORM MUSCLES
            3 CONTINUE
       C      COMPUTE LAMDAT(I1)=XAM(I1)*LAMDATO(I1) FOR ALL PENNIFORM MUSCLES, FOR THE
       C      CURRENT VALUE OF K.
              CALL FZERO(FCFIB,1.0001,1.4,F3,XA,MAXIT2,MES2)
              IF(MES2.EQ.1) WRITE(LW,20)
              XAM(I1)=XA
            4 CONTINUE
            2 TOTOR(K)=TOTOR(K)+PAR(J3)*TLX(I1)*PENFAC(I1)*DIST(I1,K)
       C      IN THE FOLLOWING STATEMENT THE DIFFERENCE (ETO(K)-TOTOR(K)) BETWEEN
       C      EXPERIMENTAL TORQUE AND MODEL PREDICTION WILL BE DIVIDED BY 100 IN ORDER
       C      TO NORMALIZE F2(K) FOR THE MINIMIZATION ROUTINE.
            1 F2(K)=W(K)*(ETO(K)-TOTOR(K))/100.
           20 FORMAT(61H NO CONVERGENCE IN FZERO OBTAINED WITH PRESENT VALUE OF
             *MAXIT)
       C
              RETURN
              END
              SUBROUTINE FZERO(FCXIX,XIA,XIB,F1,XIX,MAXIT,MES)
       C************************************************************************
       C      THIS SUBROUTINE USES THE SECANT METHOD TO COMPUTE THE ZERO OF THE FUNCTION
       C      F1(XIX). THE ROUTINE CALLS THE USER-SUPPLIED SUBROUTINE FCXIX(XIX,F1)
       C      WHICH, FOR A GIVEN VALUE OF XIX, PROVIDES THE VALUE OF F1. FCXIX MUST BE
       C      DECLARED EXTERNAL IN THE CALLING PROGRAM. XIA AND XIB DENOTE RESPECTIVELY
       C      THE LOWER AND UPPER LIMITS OF THE INITIAL GUESS OF THE INTERVAL IN WHICH
       C      F1(XIX) IS EXPECTED TO HAVE A ROOT. THE MAXIMUM NUMBER MAXIT OF ALLOWABLE
       C      ITERATIONS OF THE ALGORITHM IS SET BY THE USER (SET MAXIT=100). NORMALLY,
       C      MES=0, BUT IS SET EQUAL TO 1 IF THE NUMBER OF ITERATIONS EXCEEDS MAXIT.
       C************************************************************************
```

```
      ICOUNT=0
      MES=0
      XTEST=0.
1     ICOUNT=ICOUNT+1
      IF(ICOUNT.EQ.1) GO TO 4
      GO TO 5
4     XIO=XIA
      XI1=XIB
      CALL FCXIX(XIO,F1O)
      CALL FCXIX(XI1,F11)
5     XI2=XIO-F1O*(XI1-XIO)/(F11-F1O)
      CALL FCXIX(XI2,F12)
      RELER=ABS(XI2-XTEST)/ABS(XI2)
      IF(RELER.LT.0.001) GO TO 3
      IF(ICOUNT.GE.MAXIT) GO TO 2
      XTEST=XI2
      FTEST=F12*F1O
      IF(FTEST.GT.0.) GO TO 6
      XI1=XI2
      F11=F12
      GO TO 1
6     XIO=XI2
      F1O=F12
      GO TO 1
2     MES=1
3     XIX=XI2
      F1=F12
      RETURN
      END
      SUBROUTINE FCXIX(X,F)
C***********************************************************************
C     THIS SUBROUTINE COMPUTES THE FUNCTION WHOSE ZERO IS SOUGHT.
C***********************************************************************
      COMMON/EXFUV/ETO(80),DIST(10,80),EL(10,80),ANG(80)
      COMMON/FOCOS/I1,K,M,NPAR,TAD(10),EXSIG(10),SAL(10),ALB(10),
     *              ALSO(10),IFUS(10),THETAB(10),D(10),ALTO(10),ELB(10),
     *              FBEST(10),ELBES(10),THEMAO(10),TATHEO(10),THEMIN(10),
     *              SIGMA,LASOIN(10),SMIN
      COMMON/ESTPAR/PAR(40)
      COMMON/TLREL/TLX(10),PENFAC(10)
C
      J1=NPAR*(I1-1)+1
      J2=J1+1
      J3=J2+1
      J4=J3+1
C     IT IS ASSUMED THAT THE RESTING ACTIVE STATE IS QQO=0.005 SO THAT PENFAC=
C     1./(1.-QQO).
      PENFAC(I1)=1./0.995
      QQO=0.005
      TLX(I1)=EXP(-((X-1.)/PAR(J2))**2)
      XIKA=X
      FO=FORF(SAL(I1),EL(I1,K),ALB(I1),XIKA,0.,ALSO(I1),EXSIG(I1))
C     COMPUTE FUNCTION VALUE
      F=FO-(TLX(I1)-QQO*(2.-EXP(6.97*(X-1.))))/(1.-QQO)
      RETURN
      END
      SUBROUTINE FCFIB(XA,F)
C***********************************************************************
C     THIS SUBROUTINE COMPUTES THE FUNCTION WHOSE ZERO IS SOUGHT.
C***********************************************************************
      COMMON/EXFUV/ETO(80),DIST(10,80),EL(10,80),ANG(80)
      COMMON/FOCOS/I1,K,M,NPAR,TAD(10),EXSIG(10),SAL(10),ALB(10),
     *              ALSO(10),IFUS(10),THETAB(10),D(10),ALTO(10),ELB(10),
     *              FBEST(10),ELBES(10),THEMAO(10),TATHEO(10),THEMIN(10),
     *              SIGMA,LASOIN(10),SMIN
      COMMON/ESTPAR/PAR(40)
      COMMON/TLREL/TLX(10),PENFAC(10)
C
      J1=NPAR*(I1-1)+1
      J2=J1+1
      J3=J2+1
```

```
      J4=J3+1
      SI=SIN(THETAB(I1))
      ELTA=EL(I1,K)-XA*ALTO(I1)
      SR=SQRT(D(I1)**2+ELTA**2)
      PENFAC(I1)=ELTA/SR
      EXPA=EXP(SIGMA*(XA-1.)/PAR(J1))
      TENFO=SI*(EXPA-1.)/PENFAC(I1)
      IF(LASOIN(I1).EQ.0) GO TO 1
      ALFUN=ALSO(I1)+ALOG(1.+TENFO)/SAL(I1)
      GO TO 2
    1 ALFUN=0.
    2 XI=(SR-ALFUN)/ALB(I1)
      TLX(I1)=EXP(-((XI-1.)/PAR(J2))**2)
      QQO=0.005
      F=TENFO/EXSIG(I1)-(TLX(I1)-QQO*(2.-EXP(6.97*(XI-1.))))/(1.-QQO)
C     THE FACTOR PENFAC(I1)/(1.-QQO) IS NEEDED OUTSIDE THIS ROUTINE. WITH
C     QQO=0.005 WE COMPUTE
      PENFAC(I1)=PENFAC(I1)/0.995
      RETURN
      END
      FUNCTION FORF(SAL,EL,ALB,XI,ALBR,ALSO,EXSIG)
      FORF=(EXP(SAL*(EL-ALB*XI-ALBR-ALSO))-1.)/EXSIG
      RETURN
      END
      SUBROUTINE ANCOMP(DELANG,NDATA)
C****************************************************************
C     THIS SUBROUTINE COMPUTES THE NDATA ANGULAR POSITIONS ANG(I), I=2,...,NDATA.
C     NOTE THAT ANG(1) MUST BE READ IN BY THE USER.
C****************************************************************
      COMMON/EXFUV/ETO(80),DIST(10,80),EL(10,80),ANG(80)
      DO 1 I=2,NDATA
      K=I-1
    1 ANG(I)=ANG(K)+DELANG
      RETURN
      END
      SUBROUTINE DICOMP(M,NDATA)
C****************************************************************
C     THIS ROUTINE COMPUTES THE ELEMENTS OF THE MATRIX DIST(I,J) ACCORDING TO
C     THE EQUATION SUPPLIED BY THE USER.
C****************************************************************
      COMMON/EXFUV/ETO(80),DIST(10,80),EL(10,80),ANG(80)
C     INSERT STATEMENTS COMPUTING THE ELEMENTS OF DIST(I,J) HERE.
      RETURN
      END
      SUBROUTINE ELCOMP(M,NDATA)
C****************************************************************
C     THIS ROUTINE COMPUTES THE ELEMENTS OF ARRAY EL(I,J) ACCORDING TO THE
C     COMPUTATIONAL FORMULA SUPPLIED BY THE USER.
C****************************************************************
      COMMON/EXFUV/ETO(80),DIST(10,80),EL(10,80),ANG(80)
C     INSERT STATEMENTS COMPUTING THE ELEMENTS OF EL(I,J) HERE.
      RETURN
      END
```

## APPENDIX A4

# Computer program MYOSIM for simulating myocybernetic models

The computer program MYOSIM has been written in ANSI FORTRAN IV and simulates the four myocybernetic models described in Chapter 6. Detailed instructions on the use of the program are contained in the comment section at the beginning of the program listing, and need therefore not be repeated here. However, some remarks concerning the integration routine RKAINT may be in order.

In the routine use is made of a fourth-order Runge-Kutta-Merson integration algorithm with variable or fixed step size. In the program MYOSIM only the fixed step size version of the algorithm is used, for the following reason. Most of the differential equations describing the excitation dynamics of the four myocybernetic models contain control parameters that may change discontinuously at a given instant (so-called bang-bang controls). If the user ensures that these discontinuities occur only at multiples of the fixed integration step size HSTEP, the routine RKAINT is able appropriately to handle the discontinuities. It must, however, be stressed that more efficient integration routines are available and may be substituted for RKAINT, *provided* the user ensures that the chosen routine is capable of dealing with the type of discontinuities mentioned above, and the integration step size can be reduced by the subroutine XDTCOM, if necessary.

The following is a list of the names of the constants and variables used in the program, and also mentioned in the main text. Where text name and program name coincide, the definition will not be repeated here. The dimension of a variable or constant is not indicated.

NS = s  FRE = $\nu$  ALBEO = $\lambda_B^o$  DEL = $\delta$
ALFAB = $\bar{\alpha}$  FSEFB = $F^{SE}/\bar{F}$  EM = m  ENCO = $\hat{n}$
ALSO = $\lambda_{so}$  QQO = $q_o$  STAN = $s_k$  EDTS = $\eta^*$
ALB = $\bar{\lambda}$  SIGMA = $\sigma$  CBAR = $\bar{c}$  ALBE = $\lambda_B$
CAPPAB = $\bar{\kappa}$  XIHAT = $\hat{\xi}$  A2D = $a_2'$  EL = $\ell$
ELB = $\bar{\ell}$  A6D = $a_6'$  A3D = $a_3'$

For further explanations see the comments in the program listing.

```
C       PROGRAM MYOSIM.
C
C***********************************************************************
C       PROGRAM MYOSIM SIMULATES, FOR A VARIETY OF CONTRACTIVE MODES AND CONTROL
C       INPUTS, THE DYNAMIC BEHAVIOUR OF ONE OF SEVERAL DIFFERENT MODELS OF
C       MAMMALIAN SKELETAL MUSCLE. THE PROGRAM THUS ENABLES THE USER TO PERFORM
C       COMPUTER EXPERIMENTS ON A MUSCLE MODEL OF HIS CHOICE, AND FOR SELECTED
C       MODES OF CONTRACTION AND CONTROL INPUTS.
C
C
C       MODEL 1
C
C       THIS MODEL SIMULATES THE NORMALIZED FORCE OUTPUTS FSEFB(I), I=1,...,N,
C       OF N (MAXIMUM 10) INDIVIDUAL MUSCLE FIBRES OF A FUSIFORM MUSCLE. THE
C       NEURAL CONTROLS ARE INDIVIDUAL TRAINS OF NERVE IMPULSES OF 1 MSEC DURATION
C       OCCURRING AT TIME INTERVALS OF 1/FRE(I), I=1,...,N. THE DETAILED RESPONSES
C       TO THESE NEURAL INPUTS OF THE APPROXIMATE INDIVIDUAL ACTION POTENTIALS
C       BETA (VARIABLES Y(5*I-4), I=1,...,N), THE CORRESPONDING INTRAFILAMENTARY
C       CALCIUM CONCENTRATIONS GAMMA (VARIABLES Y(5*I-2), I=1,...,N), AND THE
C       CORRESPONDING NORMALIZED FORCE OUTPUTS FSEFB(I) I=1,...,N, FOR EACH OF
C       THE N FIBRES ARE OBSERVABLE AS OUTPUT VARIABLES. THE FORCE OUTPUT FFBTOT
C       OF THE TOTAL SYSTEM IS EQUAL TO THE NORMALIZED SUM OF THE INDIVIDUAL FIBRE
C       FORCES (I.E. FFBTOT = SUM(FSEFB(I)/N)), AND CAN ALSO BE OBSERVED.
C
C       THE VALUES OF THE CONSTANTS DEFINING THE PROPERTIES OF THE FIBRES (FOR
C       HUMAN SKELETAL MUSCLE AT 37 DEGREES CELSIUS) ARE GIVEN BY -
C               EXCITATION DYNAMICS
C               FAST FIBRE                      SLOW FIBRE
C       C1 =    2.24E4                          2.24E4
C       C2 =    2.50E5                          0.71E5
C       C3 =    8.60E10                         9.20E10
C       C4 =    1.84E4                          1.84E4
C       C5 =    7.36E6                          7.36E6
C       C6 =    10.333                          10.333
C       VN =    0.09                            0.09
C       VT =    0.05                            0.05
C       QQ0 =   0.005                           0.005
C       NQ =    2                               2
C       NS =    1                               1
C               CONTRACTION DYNAMICS
C               FAST FIBRE                      SLOW FIBRE
C       A7 =    5.94                            2.70
C               (AVERAGE FAST FIBRE WITH        (AVERAGE SLOW FIBRE WITH
C               50 MS TWITCH CONTR. TIME)       120 MS TWITCH CONTR. TIME)
C       THE REMAINING CONSTANTS ARE THE SAME FOR FAST AND SLOW FIBRES =
C       A1=0.10727021, A2=1.4085387, A3=3.2, B1=0.0050251256, B2=6.9828297,
C       A6D=6.97, XIHAT=2.90, CAPPAB=0.0306, SIGMA=1.531 .
C
C       THE VALUES OF THE FOLLOWING CONSTANTS MUST BE READ IN BY THE USER
C       (UNITS ARE METER, SECOND, NEWTON) -
C       MODEL - NUMBER OF MUSCLE MODEL TO BE SIMULATED. FOR PRESENT MODEL PUT
C               MODEL=1 .
C       IGRAPH - SELECTOR FOR GRAPHICS FACILITY. IF GRAPHIC DISPLAY IS REQUIRED
C               PUT IGRAPH=1 , OTHERWISE PUT IGRAPH=0 . IF IGRAPH=1, ADDITIONAL
C               DATA MUST BE READ IN. FOR DETAILS SEE SECTION - GRAPHIC DISPLAY
C               OF MODEL RESPONSES.
C       TSTART - STARTING TIME OF THE SIMULATION. IN GENERAL, PUT TSTART=0.0 .
C               WHEN PLOTTING GRAPHS, TSTART MUST BE PUT EQUAL TO ZERO.
C       TFIN -  TIME AT WHICH SIMULATION IS TO BE TERMINATED.
C       OUTINT - TIME INTERVAL AT WHICH OUTPUT IS REQUIRED. WHEN PLOTTING GRAPHS
C               SEE EXPLANATION OF GRAPHIC DISPLAY BELOW.
C       TAB -   FOR EXPLANATION SEE SUBROUTINE RKAINT. NORMALLY, TAB=0.0 .
C       HSTEP - INTEGRATION STEP SIZE FOR FIXED-STEP-SIZE INTEGRATION. FOR MOST
C               SIMULATIONS, A VALUE OF 1.E-4 WILL BE APPROPRIATE. IN SOME
C               SIMULATION STUDIES, HOWEVER, IT MAY HAPPEN THAT THE DIFFERENTIAL
C               SYSTEM BECOMES UNSTABLE DUE TO THE ACCUMULATION OF NUMERICAL
C               ERRORS. IN THIS CASE HSTEP MUST BE REDUCED TO 5.E-5 OR LESS.
C               THE ACCURACY OF A SOLUTION CAN ALWAYS BE CHECKED BY RUNNING THE
C               SIMULATION WITH A CERTAIN VALUE OF HSTEP AND THEN RERUNNING IT
C               WITH A VALUE OF 0.5*HSTEP . IF THE TWO SOLUTIONS DO NOT DIFFER
C               SIGNIFICANTLY, THE VALUE CHOSEN FOR HSTEP IS APPROPRIATE.
C               FOR MODELS 2 AND 3, LARGER VALUES OF HSTEP (E.G. 2.E-4) MAY
C               YIELD SUFFICIENTLY ACCURATE RESULTS. CARE MUST, HOWEVER, BE
```

```
              TAKEN WHEN DISCONTINUOUS CHANGES (SWITCHINGS OF CONTROL PARA-
              METERS, SUDDEN STRETCHES OR RELEASES) ARE INTRODUCED INTO THE
              SYSTEM. THESE MAY BE IMPLEMENTED ONLY AT MULTIPLES OF HSTEP.
     T1, T2, T3, T4, SLOPE1, SLOPE2 - FOR EXPLANATION SEE SECTION-SELECTION OF
              CONTRACTIVE MODES.
     N    -    NUMBER OF MUSCLE FIBRES.
     EL   -    INITIAL LENGTH OF THE MUSCLE . LENGTH CHANGES (STRETCHING,
              SHORTENING, ETC.) MAY BE IMPOSED ON THE MUSCLE DURING THE
              SIMULATION (SEE UNDER CONTRACTIVE MODES BELOW). IT IS, HOWEVER,
              OF EXTREME IMPORTANCE THAT THE PHYSIOLOGICAL LIMITS OF THE FIBRE
              LENGTH ARE NOT EXCEEDED. THIS IMPLIES THAT THE MINIMUM AND
              MAXIMUM PERMISSIBLE FIBRE LENGTHS ELMIN AND ELMAX ARE GIVEN BY
                   ELMIN=0.6*ELB+(0.4-0.5*ALFAB)*ALSO ,
                   ELMAX=1.7*ELB-(0.7+1.7*ALFAB)*ALSO .
              THIS LIMITATION APPLIES ONLY TO MODELS 1, 2 AND 4 , BUT NOT TO
              MODEL 3 .
     ALFAB -   N-DIMENSIONAL ARRAY OF VALUES ALPHABAR. PUT ALFAB=0.05 TO 0.11 .
     ALSO  -   N-DIMENSIONAL ARRAY OF FIBRE TENDON REST LENGTHS.
     ELB   -   N-DIMENSIONAL ARRAY OF VALUES ELBAR.
     FRE   -   N-DIMENSIONAL ARRAY OF VALUES DEFINING THE STIMULATION
              FREQUENCIES FOR EACH OF THE N FIBRES. THE USER MAY SPECIFY
              THESE FREQUENCIES AS FUNCTIONS OF TIME IN SUBROUTINE FREQU. IF
              THIS IS DONE THE VALUES OF THE FREQUENCIES AS READ IN WILL BE
              OVERWRITTEN BY THE COMPUTED VALUES FRE(I). SUBROUTINE FREQU ALSO
              MAKES IT POSSIBLE TO MODEL DELAYED ONSETS OF STIMULATION OF THE
              INDIVIDUAL FIBRES. APPROXIMATE MAXIMUM VALUES FOR FRE(I), I=1,..,N
              ARE 100 HZ FOR FAST, AND 33 HZ FOR SLOW FIBRES.
     IFF   -   N-DIMENSIONAL ARRAY OF VALUES DEFINING FIBRE I AS FAST OR SLOW.
              FOR A FAST FIBRE PUT IFF(I)=1, FOR A SLOW ONE PUT IFF(I)=0 .
     YO    -   5*N-DIMENSIONAL ARRAY OF INITIAL VALUES OF THE STATE VARIABLES.
              THE STATE VARIABLES ARE (FOR I=1,...,N) -
              Y(5*I-4) - NORMALIZED ACTION POTENTIAL BETA(I) OF THE I-TH FIBRE,
              Y(5*I-3) - TIME RATE OF CHANGE OF THE ACTION POTENTIAL,
              Y(5*I-2) - INTRAFILAMENTARY CALCIUM CONCENTRATION GAMMA(I),
              Y(5*I-1) - TIME RATE OF CHANGE OF GAMMA(I),
              Y(5*I)   - NORMALIZED LENGTH XI(I) OF THE CONTRACTILE ELEMENT OF
                         THE I-TH FIBRE.
              FOR INITIALLY RESTING MUSCLE FIBRES (WHICH IS ASSUMED HERE) PUT
                   YO(5*I-4)=0.0 ,
                   YO(5*I-3)=0.0 ,
                   YO(5*I-2)=0.0 ,
                   YO(5*I-1)=0.0 ,
                   YO(5*I)  =1.0 .
              ENTER THE 5 INITIAL STATES OF THE FIRST FIBRE FIRST, FOLLOWED BY
              THE STATES OF THE SECOND FIBRE (ON THE NEXT DATA CARD), ETC.
     BLANK CARD - MUST BE INSERTED JUST BEFORE THE END-OF-FILE CARD. IT
              PROVIDES A MACHINE-INDEPENDENT END-OF-FILE TEST. IF MORE THAN ONE
              DATA SET IS PROCESSED, THE BLANK CARD MUST FOLLOW THE LAST DATA
              CARD, JUST BEFORE THE END-OF-FILE CARD.

     SIMULATION OUTPUT -
     AT EACH TIME T(K)=TSTART+K*OUTINT, K=0,1,...., OF THE SIMULATION THE
     FOLLOWING MODEL RESPONSES ARE AVAILABLE.
     THE INDIVIDUAL NORMALIZED ACTION POTENTIALS BETA(I)=Y(5*I-4), I=1,...,N.
     INTRAFILAMENTARY CALCIUM CONCENTRATIONS GAMMA(I)=Y(5*I-2), I=1,...,N.
     NORMALIZED CONTRACTILE ELEMENT LENGTHS XI(I)=Y(5*I), I=1,...,N.
     ACTIVE STATES Q(I), I=1,...,N.

C    INDIVIDUAL NORMALIZED FIBRE FORCE OUTPUTS FSEFB(I), I=1,...,N.
C    TOTAL NORMALIZED MUSCLE FORCE FFBTOT,
C    THE LENGTH EL OF THE MUSCLE.
C
C
C
C    MODEL 2
C
C    SIMILAR TO MODEL 1, EXCEPT THAT THE CALCIUM DYNAMICS OF THE I-TH FIBRE IS
C    NOW REPRESENTED BY THE AVERAGE RESPONSE GAMMA(I), DEFINED BY
C         GAMMADOT(I) = M*(C*V(I) - GAMMA(I)) , I=1,...,N.
C    WHERE V(I) IS THE NORMALIZED STIMULATION RATE OF THE I-TH FIBRE.
C
C    ADDITIONAL VALUES OF CONSTANTS REQUIRED ARE -
C              EXCITATION DYNAMICS
C         FAST FIBRE                      SLOW FIBRE
```

```
C      M =      11.25                    3.10
C      C =      1.373E-4                 1.373E-4
C      THE CONTRACTION DYNAMICS IS UNCHANGED.
C
C      THE VALUES OF THE FOLLOWING CONSTANTS MUST BE READ IN BY THE USER
C      (UNITS ARE METER, SECOND, NEWTON) -
C      AS FOR MODEL 1, EXCEPT
C      MODEL = 2 ,
C      YO -      2*N-DIMENSIONAL ARRAY OF INITIAL VALUES OF THE STATE VARIABLES.
C                THE STATE VARIABLES (FOR I=1,...,N) ARE -
C                Y(2*I-1) - GAMMA(I),
C                Y(2*I)   - NORMALIZED LENGTH XI(I) OF THE CONTRACT. ELEMENT.
C                FOR INITIALLY RESTING MUSCLE FIBRES (WHICH IS ASSUMED HERE) PUT
C                          YO(2*I-1)=0.0 ,
C                          YO(2*I)=1.0 .
C                ENTER THE 2 INITIAL STATES OF THE FIRST FIBRE FIRST, FOLLOWED BY
C                THE STATES OF THE SECOND FIBRE (ON THE NEXT DATA CARD), ETC.
C      BLANK CARD - SEE EXPLANATION IN MODEL 1 .
C
C      SIMULATION OUTPUT -
C      AT EACH TIME T(K)=TSTART+K*OUTINT, K=0,1,..., OF THE SIMULATION THE
C      FOLLOWING MODEL RESPONSES ARE AVAILABLE.
C      INTRAFILAMENTARY CALCIUM CONCENTRATIONS GAMMA(I)=Y(2*I-1), I=1,...,N.
C      NORMALIZED CONTRACTILE ELEMENT LENGTHS XI(I)=Y(2*I), I=1,...,N.
C      ACTIVE STATES Q(I), I=1,...,N.
C      INDIVIDUAL NORMALIZED FIBRE FORCE OUTPUTS FSEFB(I), I=1,...,N.
C      TOTAL NORMALIZED MUSCLE FORCE FFBTOT,
C      THE LENGTH EL OF THE MUSCLE.
C
C
C      MODEL 3
C
C      THIS MODEL SIMULATES THE CONTROL BEHAVIOUR OF A WHOLE FUSIFORM MUSCLE
C      CONSISTING OF A LARGE NUMBER OF MOTOR UNITS, ALL OF WHICH ARE OF DIFFERENT
C      SIZE AND TYPE. THE NEURAL CONTROL INPUTS ARE THE NORMALIZED AVERAGE
C      STIMULATION RATE V (0.LE.V.LE.1), AND THE NORMALIZED RECRUITMENT RATE Z
C      (-ZMIN.LE.Z.LE.1). NOTE THAT Z MAY NOT ALTERNATE IN SIGN TOO RAPIDLY.
C
C      THE AVERAGE VALUES OF THE CONSTANTS DEFINING THE PROPERTIES OF THE MOTOR
C      UNITS (FOR HUMAN MUSCLE AT 37 DEGREES CELCIUS) ARE GIVEN BY -
C      CBAR=3.7, A2D=0.14, A3D=0.105, ENCO=14.3 . THE CONSTANT DEL HAS A VALUE OF
C      1.E-4 TO 1.E-8, WHILE THE REMAINING CONSTANTS ARE THE SAME AS FOR SINGLE
C      FIBRES. NOTE THAT THE VALUES OF A2D AND A3D CORRESPOND RESPECTIVELY TO
C      EXTREMELY SLOW (140 MS TWITCH CONTR. TIME) AND EXTREMELY FAST (35 MS
C      TWITCH CONTR. TIME) FIBRES.
C
C      THE VALUES OF THE FOLLOWING CONSTANTS MUST BE READ IN BY THE USER
C      (UNITS ARE METER, SECOND, NEWTON) -
C      MODEL = 3,
C      IGRAPH, TSTART, TFIN, ..., SLOPE2 - AS FOR MODEL 1,
C      N        = 1,
C      EL, ALFAB, ALSO, ELB - AS FOR MODEL 1,
C      YO -      5*N-DIMENSIONAL ARRAY OF INITIAL VALUES OF THE STATE VARIABLES.
C                THE STATE VARIABLES ARE -
C                Y(1) - NORMALIZED POPULATION EN OF STIMULATED MOTOR UNITS,
C                Y(2) - NORMALIZED POPULATION R OF SEMI-ACTIVE MOTOR UNITS,
C                Y(3) - PSEUDO-CALCIUM CONCENTRATION PSI OF THE EN-POPULATION,
C                Y(4) - PSEUDO-CALCIUM CONCENTRATION PHI OF THE R-POPULATION,
C                Y(5) - NORMALIZED LENGTH XI OF CONTRACTILE ELEMENT.
C      FOR INITIALLY RESTING MUSCLE (WHICH IS ASSUMED HERE) PUT
C                          YO(1) = 0.0
C                          YO(2) = 0.0
C                          YO(3) = 0.0
C                          YO(4) = 0.0
C                          YO(5) = 1.0
C      BLANK CARD - SEE EXPLANATION IN MODEL 1 .
C      THE CONTROL FUNCTIONS V(T) AND Z(T) MUST ALSO BE SPECIFIED BY THE USER IN
C      THE SUBROUTINE CONTRL .
C
C      SIMULATION OUTPUT -
C      AT EACH TIME T(K)=TSTART+K*OUTINT, K=0,1,..., OF THE SIMULATION THE
C      FOLLOWING MODEL RESPONSES ARE AVAILABLE.
C      NORMALIZED POPULATION EN=Y(1) OF STIMULATED MOTOR UNITS,
C      NORMALIZED POPULATION R=Y(2) OF SEMI-ACTIVE MOTOR UNITS,
```

```
C     NORMALIZED LENGTH XI=Y(5) OF THE CONTRACTILE ELEMENT,
C     ACTIVE STATE Q(EN)=Q(1) OF THE EN-POPULATION,
C     ACTIVE STATE Q(R)=Q(2) OF THE R-POPULATION,
C     TOTAL NORMALIZED MUSCLE FORCE OUTPUT FFBTOT,
C     THE LENGTH EL OF THE MUSCLE.
C
C
C     MODEL 4
C
C     THIS MODEL SIMULATES THE BEHAVIOUR OF A NUMBER (MAXIMUM 10) OF INDEPENDENT
C     MOTOR UNITS OF EXPONENTIALLY INCREASING SIZES.
C     THE SPECIFICATIONS FOR THIS MODEL ARE PRECISELY THE SAME AS FOR MODEL 1,
C     EXCEPT THAT MODEL=4 .
C
C
C     SELECTION OF MODE OF CONTRACTION
C
C     THE PROGRAM ALLOWS THE USER TO SELECT NOT ONLY THE CONTROL MODE (STIMUL-
C     ATION RATES, RECRUITMENT RATES) BUT ALSO THE CONTRACTIVE MODE OF THE
C     MUSCLE (FIBRE) BY VARYING THE MUSCLE LENGTH EL(T) IN A SPECIFIC WAY.
C     THE SUBROUTINE LENGTH PERFORMS THIS FUNCTION AS FOLLOWS.
C     FROM TIME TSTART TO TIME T1 THE MUSCLE (FIBRE) IS KEPT AT THE VALUE EL
C     OF THE LENGTH WHICH WAS READ IN BY THE USER. FROM T1 TO T2 THE MUSCLE IS
C     LENGTHENED (OR SHORTENED) AT A RATE SLOPE1 (SHORTENING TAKES PLACE FOR
C     SLOPE1 NEGATIVE). FROM T2 TO T3 THE LENGTH IS AGAIN KEPT CONSTANT AT THE
C     LAST VALUE, WHILE FROM T3 TO T4 A FURTHER LENGTH CHANGE AT RATE SLOPE2 CAN
C     BE PERFORMED. AFTER T4, THE LENGTH AGAIN REMAINS CONSTANT AT EL(T4).
C     BY SPECIFYING THE VALUES OF T1,T2,T3,T4,SLOPE1,SLOPE2, THE USER CAN
C     PRODUCE A SEQUENCE OF CONTRACTIVE MODES. IN ADDITION, THE USER CAN DESIGN
C     HIS OWN FUNCTION EL(T) IN SUBROUTINE LENGTH.
C
C
C     GRAPHIC DISPLAY OF MODEL RESPONSES
C
C     IF A CALCOMP FACILITY IS AVAILABLE AT THE USER'S COMPUTER INSTALLATION,
C     ALL MODEL RESPONSES SIMULATED BY THE PROGRAM CAN BE DISPLAYED GRAPHICALLY
C     ON AN ELECTRONIC SCREEN, OR IN THE FORM OF A PLOT. TO OBTAIN A GRAPHIC
C     DISPLAY PUT IGRAPH=1 . A MAXIMUM OF 5 PLOTS CAN BE DISPLAYED SIMULTANEOUS-
C     LY, INVOLVING ANY OF THE OUTPUT VARIABLES. ALL VARIABLES ARE NORMALIZED
C     FOR THIS PURPOSE. IT IS ALSO POSSIBLE TO DISPLAY THE MODEL RESPONSES OF
C     DIFFERENT SIMULATION RUNS (MAXIMUM=3) ON A SINGLE GRAPH. IN THIS CASE,
      HOWEVER, THE USER MUST SUPPLY THE DIFFERENT DATA SETS THAT DEFINE THE
      INPUT DATA FOR THE SUCCESSIVE RUNS IN THE TOTAL INPUT DATA SET. FOR
      EXAMPLE, IF 2 SUCCESSIVE RUNS ARE TO BE EXECUTED, THE COMPLETE DATA SET
      DEFINING THE INPUT FOR THE FIRST RUN IS FOLLOWED BY THE COMPLETE DATA SET
      WHICH DEFINES THE INPUT FOR THE SECOND RUN. NO OTHER RECORDS NEED BE
      INSERTED IN BETWEEN. THE END OF THE TOTAL DATA RECORD IS INDICATED BY THE
      BLANK CARD INSERTED JUST BEFORE THE END-OF-FILE CARD. NOTE, HOWEVER, THAT
      THE OUTPUTS OF REPEATED RUNS WHICH ARE TO BE DISPLAYED ON THE SAME GRAPH
      MUST ORIGINATE FROM THE SAME MODEL.
      THE FOLLOWING PROGRAM CONTROL VARIABLES MUST BE SPECIFIED BY THE USER
      (THEIR VALUES HAVE TO BE READ IN ONLY IF IGRAPH=1 ) -
      IFIB -   NUMBER OF THE FIBRE (FOR MODELS 1 AND 2) OR MOTOR UNIT (FOR
               MODEL 4) WHOSE OUTPUT VARIABLES SHOULD BE DISPLAYED. NOTE THAT
               1.LE.IFIB.LE.N , WHERE MAX(N)=10 .
      LINE -   INTEGER VARIABLE PERMITTING SELECTION OF LINE DRAWING MODE. IF
               LINE=1, ALL CURVES ON A GRAPH WILL BE REPRESENTED BY CONTINUOUS
               LINES. IF LINE=0, ONLY THE FIRST CURVE WILL BE CONTINUOUS, ALL
               SUBSEQUENT CURVES ON THIS GRAPH WILL CONSIST OF DOTTED LINES.
      NPLOTS - NUMBER OF PLOTS TO BE DISPLAYED ON A SINGLE GRAPH (MAXIMUM=5),
               PER SIMULATION RUN. NOTE THAT FOR MULTIPLE RUNS THE NUMBER NPLOTS
               DESIGNATES THE NUMBER OF PLOTS PER RUN. HOWEVER, ALL PLOTS WILL
               BE COMBINED TO A SINGLE GRAPH DISPLAYING THE RESULTS OF ALL THE
               SIMULATION RUNS.
      INDEX(1), INDEX(2), ...., INDEX(NPLOTS) - INTEGER VALUES DESIGNATING THE
               OUTPUT VARIABLES TO BE DISPLAYED. THE VALUES ARE DEFINED BY -
               FOR MODELS 1 AND 4 -
                 BETA(IFIB)  GAMMA(IFIB)  XI(IFIB)  Q(IFIB)  FSEFB(IFIB)  FFBTOT  EL
               INDEX(.)= 1       2            3        4         5          6     7
                 FOR MODEL 2 -
                 GAMMA(IFIB)  XI(IFIB)  Q(IFIB)  FSEFB(IFIB)  FFBTOT  EL
               INDEX(.)= 1        2         3         4          5     6
                 FOR MODEL 3 -
```

```
           EN   R    XI   Q(EN)   Q(R)   FFBTOT   EL
INDEX(.)=1  2   3    4     5       6       7
   IF, FOR EXAMPLE, A SIMULTANEOUS DISPLAY OF THE VARIABLES Q, FFBTOT, AND
   EL OF MODEL 2 IS REQUIRED, THE USER PUTS NPLOTS=3, INDEX(1)=3,
   INDEX(2)=5, AND INDEX(3)=6 .
A MAXIMUM OF 500 POINTS PER PLOT IS PERMITTED. THE CORRESPONDING
PERMISSIBLE MINIMUM VALUE OF THE VARIABLE OUTINT IS GIVEN BY
              OUTINT = (TFIN - TSTART)/500 ,
WHERE ALWAYS TSTART=0.0 . THIS ADJUSTMENT OF OUTINT IS PERFORMED
AUTOMATICALLY BY THE PROGRAM IF IGRAPH=1 .


ALLOCATION OF INPUT AND OUTPUT TAPE NUMBERS

IN ORDER TO KEEP THE PROGRAM AS MACHINE INDEPENDENT AS POSSIBLE, THE TAPE
NUMBERS DEFINING THE INPUT (READ) FILE LR, THE OUTPUT (WRITE) FILE LW,
AND THE PLOTTING FILES IOUT1 AND IOUTP, CAN BE SPECIFIED BY THE USER IN A
DATA STATEMENT IN THE MAIN PROGRAM. PRESENT DEFAULT VALUES ARE LR=5,
LW=6, IOUT1=8, IOUTP=10 .


REMARKS

THE VARIABLE ICBEL IN THE MAIN PROGRAM MAKES IT POSSIBLE FOR THE USER
TO RUN SIMULATIONS ASSUMING INDEPENDENCE OF INTERFILAMENTARY VELOCITY OF
THE NUMBER OF CROSS-BRIDGES ATTACHED TO THE ACTIN COMPLEX. IF SUCH A
STUDY IS DESIRED THE PRESENT VALUE OF ICBEL=0 MUST BE CHANGED TO ICBEL=1 .


EXTERNAL SUBROUTINES CALLED - THE CALCOMP ROUTINES PLOTS, PLOT, AXIS,
          AND SYMBOL. THESE ROUTINES ARE REQUIRED ONLY IF USE IS MADE OF
          THE CALCOMP GRAPHIC-DISPLAY FACILITY.
C     PROGRAM DEVELOPED BY DR. H. HATZE AND J. GEYER, PRETORIA, 1980.
C
C**********************************************************************
      DIMENSION YP(50),TOL(50),EREST(50),YY(50),CON1(50),CON3(50),
     *CON4(50),CON5(50),YG(50),Y(50)
      COMMON/ICONST/NQ,NS,N,MODEL,ICBEL
      COMMON/CONST/A1(2),A2(2),A3(2),A7(2),ALFAB(10),ALSO(10),ALBT(10),
     *B1(2),B2(2),C1(2),C2(2),C3(2),C4,C5,C6,CAPPAB,CF1(10),CF2(10),
     *ELB(10),EXSIG,FRE(10),FSEFB(10),IFF(10),PI,QQ0(2),SIGMA,VN,VT,
     *XIMAT,A6D,ALBEO(10),EM(2),C,STAN,CNS,CBAR,A2D,A3D,DEL,ENCO,EXCO,
     *EDTS(2),ALBE(10),Q(10)
      COMMON/ELCOMP/T1,T2,T3,T4,SLOPE1,SLOPE2,SLOPE,EL
      COMMON/TIME/TSTART,TFIN,HSTEP
      COMMON/INTEGR/IFBR,IFAST
      COMMON/CGRAPH/IGRAPH,IFIB,NPLOTS,INDEX(5),IRUN,LINE
      COMMON/CIOF/LR,LW,IOUT1,IOUTP
      EXTERNAL F,FCXIX
C
C     DEFINE INTERNAL FUNCTIONS
      AFU(ARG,A2)=-ALOG(ARG)/A2
      ASH(A7,A2,A3,ARG)=A7*(ALOG(AFU(ARG,A2)+SQRT(1.+AFU(ARG,A2)**2))/
     *                 A3-0.5)
C
C     IOEND=1 INDICATES THAT END OF ALL INPUT DATA RECORDS HAS BEEN REACHED.
C     OTHERWISE -
      IOEND=0
C     IRUN IS THE NUMBER OF DIFFERENT INPUT DATA SETS AND EQUALS THE NUMBER OF
C     DIFFERENT SIMULATION RUNS. ITS INITIAL VALUE IS -
      IRUN=0
  27  CONTINUE
C
C     READ ALL INPUT DATA.
C
      CALL IOFILE(NEQN,OUTINT,TAB,YO,IOEND)
      IF(IGRAPH.EQ.1)OUTINT=(TFIN-TSTART)/500.0
      IF(IOEND.EQ.1) GO TO 26
      IRUN=IRUN+1
C
C     INITIAL COMPUTATIONS OF CONSTANTS
C
      ELIN=EL
```

```
      PI=ATAN2(0.,-1.)
      EXSIG=EXP(SIGMA)-1.
      STAN=0.46
      CNS=XIHAT**NS-1.
      EXCB=EXP(CBAR)-1.
      ARG1=1./(0.999/A1(1)+0.001*B1(1)*B2(1))-A1(1)
      ARG2=1./(0.999/A1(2)+0.001*B1(2)*B2(2))-A1(2)
      EDTS(1)=ASH(1.,A2(1),A3(1),ARG1)
      EDTS(2)=ASH(1.,A2(2),A3(2),ARG2)
      MAXIT1=150
      IF(ICBEL.EQ.0) CAPPAB=0.0
C
C              BRANCHING TO THE VARIOUS MODELS
C
      GO TO (1,2,3,4),MODEL
C
C     COMPUTATIONS FOR MODELS 1, 2 AND 4
C
    1 CONTINUE
    2 CONTINUE
    4 CONTINUE
C
C     INITIAL COMPUTATIONS FOR EACH OF THE N FIBRES.
C
      DO 5 I=1,N
      IFBR=I
      IXI=I*5
      IF(MODEL.EQ.2)IXI=I*2
      ALB(I)=(ELB(I)-(1.+ALFAB(I))*ALSO(I))/(1.+CAPPAB)
      CF1(I)=SIGMA/(ALFAB(I)*ALSO(I))
      CF2(I)=0.0
      IF(ICBEL.EQ.1)CF2(I)=SIGMA/(CAPPAB*ALB(I))
      ALBEO(I)=0.0
      IF(ICBEL.EQ.1)ALBEO(I)=ALOG(1.+EXSIG/(1.-QQO(1)))/CF2(I)
      Q(I)=QQO(I)
C     FOR A FAST FIBRE IFAST=1 ELSE IFAST=2.
      IFAST=1
      IF(IFF(I).EQ.0)IFAST=2
C
C     IT IS ASUMED THAT THE FIBRE IS AT REST AT T=TSTART, AND THAT THE
C     MUSCLE IS OF FUSIFORM.
C
      ELTEST=ALB(I)+ALSO(I)
      IF(ICBEL.EQ.1)ELTEST=ALB(I)*0.97061+ALBEO(I)+ALSO(I)
      IF(EL.LT.ELTEST)GO TO 8
      XIAV=(EL-ALSO(IFBR))/ALB(IFBR)-CAPPAB
      XIDEL=1.2*ALFAB(I)*ALSO(IFBR)*(EL/ELTEST-0.7)/ALB(IFBR)
      XIA=XIAV-XIDEL
      XIB=XIAV+XIDEL/100.
      CALL FZERO(FCXIX,XIA,XIB,F1,YO(IXI),MAXIT1,MES1)
      IF(MES1.NE.0)WRITE(LW,290)
      GO TO 9
    8 YO(IXI)=1.
      IF(ICBEL.EQ.1)YO(IXI)=0.97061
      FSEFB(I)=0.0
    9 CONTINUE
    5 CONTINUE
C
C     INITIALIZE STATE VARIABLES
C
      DO 11 I=1,NEQN
   11 Y(I)=YO(I)
C
C     SET PARAMETERS FOR INTEGRATION
C
      T=TSTART
      FLAG=0.
      IABSER=-1
      GO TO 6
    7 CONTINUE
      H=HSTEP
      CALL RKAINT(Y,YP,T,NEQN,F,IABSER,H,HINTV,OUTINT,TOL,TAB,INDIC,
     *EREST,FLAG,YY,CON1,CON3,CON4,CON5)
```

```
      6 FFBTOT=0.
        DO 13 I=1,N
C       THE FOLLOWING CALL TO XDTCOM IS FOR THE COMPUTATION OF FSEFB(I) AND Q(I),
C       USING THE NEW STATE Y(T).
        J=1
        IF(IFF(I).EQ.0)J=2
        I5=I*5
        I3=I5-2
        IF(MODEL.EQ.2)I5=I*2
        IF(MODEL.EQ.2)I3=I5-1
        CALL XDTCOM(Y(I5),Y(I3),T,TSTART,NQ,XIHAT,NS,QQO(J),ALBE(I),EXSIG,
       *CF1(I),CF2(I),ALB(I),ALSO(I),A6D,B1(J),B2(J),FSEFB(I),A1(J),
       *A2(J),A3(J),A7(J),YP(I5),HINTV,INDIC,STAN,MODEL,CNS,EDTS(J),Q(I),
       *0,ICBEL)
        IF(MODEL.EQ.4) FFBTOT=FFBTOT+FSEFB(I)*(EXP(CBAR*FLOAT(I)/FLOAT(N))
       *                       -EXP(CBAR*FLOAT(I-1)/FLOAT(N)))/EXCB
     13 IF(MODEL.LT.4)FFBTOT=FFBTOT+FSEFB(I)/FLOAT(N)
        IF(MODEL.EQ.2)GO TO 14
        DO 15 K=1,N
        K1=K*5-4
        K3=K*5-2
        K5=K*5
        IF(IGRAPH.EQ.1.AND.K.EQ.IFIB)WRITE(IOUT1,85)T,Y(K1),Y(K3),Y(K5),
       *                             Q(K),FSEFB(K),FFBTOT,EL
     15 WRITE(LW,70)T,Y(K1),Y(K3),Y(K5),Q(K),FSEFB(K),FFBTOT,EL
        GO TO 16
     14 CONTINUE
        DO 12 K=1,N
        K1=K*2-1
        K2=K*2
        IF(IGRAPH.EQ.1.AND.K.EQ.IFIB)WRITE(IOUT1,85)T,Y(K1),Y(K2),Q(K),
       *                             FSEFB(K),FFBTOT,EL
     12 WRITE(LW,75)T,Y(K1),Y(K2),Q(K),FSEFB(K),FFBTOT,EL
     16 IF(T.GE.(TFIN-1.E-8))GO TO 27
        GO TO 7
C
C       COMPUTATIONS FOR MODEL 3
C
      3 CONTINUE
        ALB(1)=(ELB(1)-(1.+ALFAB(1))*ALSO(1))/(1.+CAPPAB)
        CF1(1)=SIGMA/(ALFAB(1)*ALSO(1))
        CF2(1)=0.0
        IF(ICBEL.EQ.1)CF2(1)=SIGMA/(CAPPAB*ALB(1))
        ALBEO(1)=0.0
        IF(ICBEL.EQ.1)ALBEO(1)=ALOG(1.+EXSIG/(1.-QQO(1)))/CF2(1)
C       IT IS ASSUMED THAT THE MUSCLE IS INITIALLY AT REST
        IFAST=1
        IFBR=1
        ELTEST=ALB(1)+ALSO(1)
        IF(ICBEL.EQ.1)ELTEST=ALB(1)*0.97061+ALBEO(1)+ALSO(1)
        IF(EL.LT.ELTEST)GO TO 17
        XIAV=(EL-ALSO(IFBR))/ALB(IFBR)-CAPPAB
        XIDEL=1.2*ALFAB(1)*ALSO(1)*(EL/ELTEST-0.7)/ALB(1)
        XIA=XIAV-XIDEL
        XIB=XIAV+XIDEL/100.
        CALL FZERO(FCXIX,XIA,XIB,F1,YO(5),MAXIT1,MES1)
        IF(MES1.NE.0)WRITE(LW,290)
        GO TO 18
     17 YO(5)=1.
        IF(ICBEL.EQ.1)YO(5)=0.97061
        FSEFB(1)=0.0
     18 FFBTOT=FSEFB(1)
C
C       INITIALIZE STATE VARIABLES
        DO 19 I=1,NEQN
     19 Y(I)=YO(I)
C
        T=TSTART
        FLAG=0.
        IABSER=-1
        IF(IGRAPH.EQ.1)WRITE(IOUT1,85)T,YO(1),YO(2),YO(5),QQO(1),QQO(1),
       *                FFBTOT,EL
        WRITE(LW,80)T,YO(1),YO(2),YO(5),QQO(1),QQO(1),FFBTOT,EL
```

```
   20 CONTINUE
      H=HSTEP
      CALL RKAINT(Y,YP,T,NEQN,F,IABSER,H,HINTV,OUTINT,TOL,TAB,INDIC,
     *EREST,FLAG,YY,CON1,CON3,CON4,CON5)
C     THE FOLLOWING CALL TO XDTCOM IS FOR THE COMPUTATION OF FFBTOT, USING THE
C     NEW STATE Y(T).
      CALL EPSILN(CBAR,Y(1),Y(2),EXCB,NQ,CNS,XIHAT,NS,QQO(1),Y(3),Y(4),
     *            Q(1),Q(2),Y(5),EPSN,EPSR,EPSO,EPSI)

      CALL XDTCOM(Y(5),EPSI,T,TSTART,NQ,XIHAT,NS,QQO(1),ALBE(1),EXSIG,
     *            CF1(1),CF2(1),ALB(1),ALSO(1),A6D,B1(1),B2(1),FSEFB(1),
     *            A1(1),A2(1),A3(1),1.,YP(5),HINTV,INDIC,STAN,MODEL,
     *            CNS,EDTS(1),EPS,0,ICBEL)
C
      FFBTOT=FSEFB(1)
      IF(IGRAPH.EQ.1)WRITE(IOUT1,85)T,Y(1),Y(2),Y(5),Q(1),Q(2),
     *                              FFBTOT,EL
      WRITE(LW,80)T,Y(1),Y(2),Y(5),Q(1),Q(2),FFBTOT,EL
      IF(T.GE.(TFIN-1.E-8))GO TO 27
      GO TO 20
   26 IF(IGRAPH.EQ.0)STOP
      REWIND IOUT1
      CALL PLOTR(OUTINT,ELIN)
      STOP
C
   70 FORMAT(F10.4,5X,2(E13.6,5X),4(F6.4,5X),F10.6)
   75 FORMAT(F10.4,5X,E13.6,5X,4(F6.4,5X),F10.6)
   80 FORMAT(F10.4,5X,6(F6.4,5X),F10.6)
   85 FORMAT(8E13.6)
  290 FORMAT(5X,62HMAXIMUM NUMBER MAXIT1 OF ITERATIONS IN SUBROUT. FZERO
     * EXCEEDED)
      END
      BLOCK DATA
C***********************************************************************
C     INITIALIZE CONSTANTS IN LABELED COMMON BLOCKS
C***********************************************************************
      COMMON/ICONST/NQ,NS,N,MODEL,ICBEL
      COMMON/CONST/A1(2),A2(2),A3(2),A7(2),ALFAB(10),ALSO(10),ALB(10),
     *B1(2),B2(2),C1(2),C2(2),C3(2),C4,C5,C6,CAPPAB,CF1(10),CF2(10),
     *ELB(10),EXSIG,FRE(10),FSEFB(10),IFF(10),PI,QQO(2),SIGMA,VN,VT,
     *XIHAT,A6D,ALBEO(10),EM(2),C,STAN,CNS,CBAR,A2D,A3D,DEL,ENCO,EXCB,
     *EDTS(2),ALBE(10),Q(10)
      COMMON/CIOF/LR,LW,IOUT1,IOUTP
C
      DATA LR,LW,IOUT1,IOUTP,ICBEL/5,6,8,10,0/
C
C     DATA FOR MODELS 1, 2 AND 4
C
      DATA C1(1),C1(2),C2(1),C2(2),C3(1)/2*2.24E4,2.5E5,0.71E5,8.6E10/,
     *     C3(2),C4,C5,C6,VN,VT/9.2E10,1.84E4,7.36E6,10.333,0.09,0.05/,
     *     C,QQO(1),QQO(2),NQ,NS,A1(1)/1.373E-4,2*0.005,2,1,0.10727021/,
     *     A1(2),A2(1),A2(2)/0.10727021,2*1.4085387/,
     *     A3(1),A3(2),A7(1),A7(2)/2*3.2,5.94,2.70/,
     *     B1(1),B1(2),B2(1),B2(2)/2*0.0050251256,2*6.9828297/,
     *     A6D,XIHAT,CAPPAB,SIGMA/6.97,2.90,0.0306,1.531/,
     *     EM(1),EM(2)/11.25,3.10/
C
C     ADDITIONAL DATA NEEDED FOR MODEL 3
C
      DATA CBAR,A2D,A3D,DEL,ENCO/3.7,0.14,0.105,1.E-4,14.3/
      END
      SUBROUTINE IOFILE(NEQN,OUTINT,TAB,YO,IOEND)
C***********************************************************************
C     THIS SUBROUTINE READS ALL INPUT DATA AND WRITES THEM OUT AFTERWARDS.
C***********************************************************************
      DIMENSION YO(50)
      COMMON/ICONST/NQ,NS,N,MODEL,ICBEL
      COMMON/CONST/A1(2),A2(2),A3(2),A7(2),ALFAB(10),ALSO(10),ALB(10),
     *B1(2),B2(2),C1(2),C2(2),C3(2),C4,C5,C6,CAPPAB,CF1(10),CF2(10),
     *ELB(10),EXSIG,FRE(10),FSEFB(10),IFF(10),PI,QQO(2),SIGMA,VN,VT,
     *XIHAT,A6D,ALBEO(10),EM(2),C,STAN,CNS,CBAR,A2D,A3D,DEL,ENCO,EXCB,
     *EDTS(2),ALBE(10),Q(10)
      COMMON/ELCOMP/T1,T2,T3,T4,SLOPE1,SLOPE2,SLOPE,EL
      COMMON/TIME/TSTART,TFIN,HSTEP
```

```
      COMMON/CGRAPH/IGRAPH,IFIB,NPLOTS,INDEX(5),IRUN,LINE
      COMMON/CIOF/LR,LW,IOUT1,IOUTP
C
      READ(LR,10)NMODEL
      IF(NMODEL) 8,7,8
C
C     READ INPUT DATA WHICH ARE COMMON TO ALL MODELS
C
    8 MODEL=NMODEL
      READ(LR,10)IGRAPH
      READ(LR,30)TSTART,TFIN,OUTINT,TAB,HSTEP
      READ(LR,40)T1,T2,T3,T4,SLOPE1,SLOPE2
      READ(LR,10)N
      READ(LR,20)EL
      READ(LR,20)(ALFAB(I),I=1,N)
      READ(LR,20)(ALSO(I),I=1,N)
      READ(LR,20)(ELB(I),I=1,N)
      GO TO (1,2,3,4),MODEL
C
C     READ INPUT DATA FOR MODELS 1, 2 AND 4
    1 CONTINUE
    2 CONTINUE
    4 CONTINUE
      READ(LR,20)(FRE(I),I=1,N)
      READ(LR,50)(IFF(I),I=1,N)
      NEQPF=5
      IF(MODEL.EQ.2)NEQPF=2
      NEQN=5*N
      IF(MODEL.EQ.2)NEQN=2*N
      IF(MODEL.NE.2)READ(LR,60)(YO(I),I=1,NEQN)
      IF(MODEL.EQ.2)READ(LR,65)(YO(I),I=1,NEQN)
    5 IF(IGRAPH.EQ.0)GO TO 6
      READ(LR,10)IFIB
      READ(LR,10)LINE
      READ(LR,10)NPLOTS
      READ(LR,10)(INDEX(I),I=1,NPLOTS)

C
    6 WRITE(LW,80)
      WRITE(LW,95)
      WRITE(LW,155)MODEL
      WRITE(LW,85)IGRAPH
      WRITE(LW,105)
      WRITE(LW,125)TSTART,TFIN,OUTINT,TAB,HSTEP
      WRITE(LW,350)
      WRITE(LW,360)T1,T2,T3,T4,SLOPE1,SLOPE2
      WRITE(LW,90)
      WRITE(LW,100)N
      WRITE(LW,210)
      WRITE(LW,220)EL
      WRITE(LW,240)
      WRITE(LW,250)(ALFAB(I),I=1,N)
      WRITE(LW,260)
      WRITE(LW,250)(ALSO(I),I=1,N)
      WRITE(LW,270)
      WRITE(LW,250)(ELB(I),I=1,N)
      IF(MODEL.NE.3)WRITE(LW,130)
      IF(MODEL.NE.3)WRITE(LW,140)(FRE(I),I=1,N)
      IF(MODEL.NE.3)WRITE(LW,235)
      IF(MODEL.NE.3)WRITE(LW,225)(IFF(I),I=1,N)
      WRITE(LW,230)NEQPF
      WRITE(LW,220)(YO(I),I=1,NEQN)
      WRITE(LW,310)
      IF(MODEL.EQ.1.OR.MODEL.EQ.4)WRITE(LW,320)
      IF(MODEL.EQ.2)WRITE(LW,325)
      IF(MODEL.EQ.3)WRITE(LW,370)
```

```
      RETURN
C
C     READ INPUT DATA FOR MODEL 3
    3 CONTINUE
      NEQPF=5
      NEQN=5
      READ(LR,70) (YO(I),I=1,NEQN)
      GO TO 5
    7 IOEND=1
      RETURN
   10 FORMAT(5I2)
   20 FORMAT(8F10.4)
   30 FORMAT(4F10.5,F10.8)
   40 FORMAT(6F10.6)
   50 FORMAT(80I1)
   60 FORMAT(5F10.8)
   65 FORMAT(2F10.0)
   70 FORMAT(5F10.8)
   80 FORMAT(1H1,10X,10HINPUT DATA)
   85 FORMAT(/5X,6HIGRAPH,I4)
   90 FORMAT(/10X,1HN)
   95 FORMAT(/5X,25HPROGRAM CONTROL VARIABLES/)
  100 FORMAT(I11)
  105 FORMAT(/7X,6HTSTART,8X,4HTFIN,8X,6HOUTINT,8X,3HTAB,8X,5HHSTEP)
  125 FORMAT(5X,F10.4,3X,F10.4,3X,F10.4,3X,F10.4,3X,F10.8)
  130 FORMAT(/5X,16HFRE(I),I=1,...,N)
  140 FORMAT(5X,8F10.2)
  155 FORMAT(/5X,5HMODEL,I5)
  210 FORMAT(/10X,2HEL)
  220 FORMAT(5X,10F10.4)
  225 FORMAT(5X,10(3X,I2))
  230 FORMAT(/5X,14HYO(I),I=1,...,I2,2H*N)
  235 FORMAT(/5X,16HIFF(I),I=1,...,N)
  240 FORMAT(/5X,18HALFAB(I),I=1,...,N)
  250 FORMAT(5X,10F10.4)
  260 FORMAT(/5X,17HALSO(I),I=1,...,N)
  270 FORMAT(/5X,16HELB(I),I=1,...,N)
  310 FORMAT(1H1,10X,11HOUTPUT DATA/)
  320 FORMAT(4X,4HTIME,10X,7HBETA(I),11X,8HGAMMA(I),8X,5HXI(I),6X,4HQ(I)
     *,5X,8HFSEFB(I),5X,6HFFBTOT,10X,2HEL/)
  325 FORMAT(4X,4HTIME,11X,8HGAMMA(I),7X,5HXI(I),7X,4HQ(I),4X,8HFSEFB(I)
     *,4X,6HFFBTOT,10X,2HEL/)
  350 FORMAT(/10X,2HT1,8X,2HT2,8X,2HT3,8X,2HT4,6X,6HSLOPE1,5X,6HSLOPE2)
  360 FORMAT(5X,6F10.4)
  370 FORMAT(4X,4HTIME,9X,2HEN,9X,1HR,10X,2HXI,8X,5HQ(EN),6X,
     *4HQ(R),5X,6HFFBTOT,10X,2HEL/)
      END
      SUBROUTINE RKAINT(Q,DERIV,T,IDIM,AUX,IABSER,HIVAL,HINTV,STEPIN,
     *                  TOL,TAB,INDIC,EREST,FLAG,Y,CON1,CON3,CON4,CON5)
C***************************************************************
C     THIS ROUTINE INTEGRATES AN IDIM-DIMENSIONAL SYSTEM
C                 QDOT=F(Q,T) ,  Q(T=0)=Q0 ,
C     OF FIRST-ORDER DIFFERENTIAL EQUATIONS. IT USES A RUNGE-KUTTA-MERSON
C     INTEGRATION ALGORITHM WITH VARIABLE OR FIXED STEP SIZE.
C
C     PARAMETERS (INPUT AND OUTPUT)
C
C     Q - IDIM-DIMENSIONAL VECTOR OF STATE VARIABLES. ON INPUT, Q CONTAINS THE
C         INITIAL VALUES. ON OUTPUT, Q CONTAINS THE SOLUTION VECTOR.
C     DERIV - IDIM-DIMENSIONAL VECTOR CONTAINING THE RIGHT HAND SIDES OF THE
C             DIFFERENTIAL SYSTEM.
C     T - INDEPENDENT VARIABLE. ON INPUT, T IS THE INITIAL TIME. ON OUTPUT, T IS
C         INCREMENTED EVERY TIME BY STEPIN; I.E. T IS THE CURRENT VALUE OF THE
C         INDEPENDENT VARIABLE. THE USER MUST SPECIFY THE VALUE OF THE INITIAL
C         TIME (USUALLY T=0.) BEFORE THE STATEMENT CALLING RKAINT.
C
C     IDIM - DIMENSION OF THE VECTOR OF STATE VARIABLES.
C     AUX - USER-SUPPLIED EXTERNAL SUBROUTINE WHICH COMPUTES THE RIGHT-HAND
C           SIDES OF THE DIFFERENTIAL SYSTEM. AUX MUST BE DECLARED EXTERNAL IN
C           THE CALLING PROGRAM AND MUST HAVE THE FORM
C           SUBROUTINE AUX(IDIM,DERIV,Q,T,HINTV,INDIC).
C     IABSER - AN INTEGER VALUE CONTROLLING ERROR AND STEP SIZE OPTIONS. PUT
C              IABSER=-1 IF A FIXED STEP SIZE HIVAL IS TO BE USED IN THE
C              INTEGRATION. IN THIS CASE, HIVAL MUST BE GIVEN A VALUE BY THE
```

```
C                    USER. IF THE INTEGRATION STEP SIZE HINTV IS TO BE ADJUSTED BY THE
C                    PROGRAM TO SATISFY AN ALLOWABLE ABSOLUTE ERROR, PUT IABSER=1. IF
C                    HINTV IS TO BE ADJUSTED FOR AN ALLOWABLE RELATIVE ERROR, PUT
C                    IABSER=0.
C           HIVAL  - INTEGRATION STEP SIZE TO BE SPECIFIED BY THE USER FOR FIXED STEP
C                    SIZE INTEGRATION (PUT IABSER=-1).
C           HINTV  - INTEGRATION INTERVAL FOR VARIABLE STEP SIZE INTEGRATION (IABSER=0
C                    OR 1). HINTV IS AUTOMATICALLY ADJUSTED BY THE ROUTINE AND ONLY AN
C                    INITIAL ESTIMATE MUST BE SPECIFIED BY THE USER. IF IABSER=-1, NO
C                    INITIAL ESTIMATE OF HINTV NEED BE PROVIDED.
C           STEPIN - INTERVAL AT WHICH OUTPUT VALUES OF Q ARE REQUIRED. STEPIN MUST BE
C                    SPECIFIED BY THE USER.
C           TOL    - IDIM-DIMENSIONAL VECTOR OF ABSOLUTE OR RELATIVE ERROR TOLERANCES TO
C                    BE SUPPLIED BY THE USER FOR EACH OF THE IDIM STATE VARIABLES.
C           TAB    - USER-SUPPLIED VALUE TO PERMIT THE PRINTING OF OUTPUT VALUES AT EACH
C                    INTEGRATION STEP. IF TAB=0., OUTPUT WILL BE PRINTED ONLY AT
C                    INTERVALS OF SIZE STEPIN. IF TAB=-1., OUTPUT WILL BE AFTER EACH
C                    INTEGRATION STEP OF SIZE HINTV.
C           INDIC  - AN INDICATOR VARIABLE WHICH IS ZERO ONLY ON ENTRY AND EXIT.
C                    DURING THE PROCESS OF INTEGRATION, INDIC IS GREATER THAN ZERO.
C           EREST  - IDIM-DIMENSIONAL VECTOR OF ERROR ESTIMATES, FOR EACH OF THE IDIM
C                    STATE VARIABLES.
C           FLAG   - INTERNAL INDICATOR INDICATING TO THE ROUTINE THAT OUTPUT IS DUE.
C                    PUT FLAG=0. BEFORE THE STATEMENT CALLING RKAINT.
C           Y,CON1,CON3,CON4,CON5 - WORKSPACES OF DIMENSION IDIM EACH.
C***********************************************************************
      DIMENSION Q(IDIM),EREST(IDIM),CON1(IDIM),CON3(IDIM),CON4(IDIM),
     *          CON5(IDIM),Y(IDIM),DERIV(IDIM),TOL(IDIM)
   12 CONTINUE
      JADIC=0
    7 CONTINUE
      INDIC=0
      IF(IABSER.EQ.-1) HINTV=HIVAL
      IF(HINTV.GT.STEPIN) HINTV=STEPIN
C
C     THE FOLLOWING CALL TO AUX SERVES TO ADJUST, WITHIN AUX, THE INTEGRATION
C     STEP HINTV BEFORE THE ACTUAL INTEGRATION COMMENCES. SUCH AN ADJUSTMENT
C     MAY BE NECESSARY DUE TO THE SPECIAL NATURE OF THE DIFFERENTIAL SYSTEM.
      CALL AUX(IDIM,DERIV,Q,T,HINTV,INDIC)
C
      IF(TAB.LT.-0.5) GO TO 11
      IF(FLAG.GT.0.5) GO TO 15
      VAL=T-FLOAT(INT(T/STEPIN))*STEPIN
      VALDIF=STEPIN-VAL
      IF(HINTV.LT.(VALDIF-1.E-6)) GO TO 11
      HINTV=VALDIF
      FLAG=1.
      GO TO 11
   15 FLAG=0.
      IF(ABS(STEPIN-HINTV).LT.1.E-9) FLAG=1.
   11 CONTINUE
      INDIC=INDIC+1
      CALL AUX(IDIM,DERIV,Q,T,HINTV,INDIC)
      DO 1 I=1,IDIM
      CON1(I)=HINTV*DERIV(I)
    1 Y(I)=Q(I)+CON1(I)/3.
      TY=T+HINTV/3.

      INDIC=INDIC+1
      CALL AUX(IDIM,DERIV,Y,TY,HINTV,INDIC)
      DO 2 I=1,IDIM
      CON3(I)=HINTV*DERIV(I)
    2 Y(I)=Q(I)+(CON1(I)+CON3(I))/6.
C     THE PREVIOUS COEFFICIENT CON3(I) ACTUALLY CORRESPONDS TO K2 OF
C     THE RUNGE-KUTTA-MERSON FORMULA.HOWEVER SINCE K2 IS NOT USED
C     AFTER THE PREVIOUS DO-LOOP,STORAGE SPACE IS SAVED BY DEFINING
C     K2 FIRST AS K2=CON3 AND THEN OVERWRITING IT WITH THE PROPER
C     K3=CON3.
      INDIC=INDIC+1
      CALL AUX(IDIM,DERIV,Y,TY,HINTV,INDIC)
      DO 3 I=1,IDIM
      CON3(I)=HINTV*DERIV(I)
    3 Y(I)=Q(I)+(CON1(I)+3.*CON3(I))/8.
      TY=T+HINTV/2.
```

```
      INDIC=INDIC+1
      CALL AUX(IDIM,DERIV,Y,TY,HINTV,INDIC)
      DO 4 I=1,IDIM
      CON4(I)=HINTV*DERIV(I)
    4 Y(I)=Q(I)+CON1(I)/2.-1.5*CON3(I)+2.*CON4(I)
      TY=T+HINTV
      INDIC=INDIC+1
      CALL AUX(IDIM,DERIV,Y,TY,HINTV,INDIC)
      DO 5 I=1,IDIM
      CON5(I)=HINTV*DERIV(I)
    5 Y(I)=Q(I)+(CON1(I)+4.*CON4(I)+CON5(I))/6.
      IF(IABSER.EQ.-1) GO TO 13
      ESTMAX=0.
      ADVAR=1.E+12
      DO 14 I=1,IDIM
      EREST(I)=ABS((2.*CON1(I)-9.*CON3(I)+8.*CON4(I)-CON5(I))/30.)
      ESTMAX=AMAX1(EREST(I),ESTMAX)
      IF(IABSER.EQ.1) ABREL=TOL(I)
      IF(IABSER.EQ.0) ABREL=TOL(I)*ABS(Y(I))
      ADRAT=ABREL/(EREST(I)+1.E-12)
      IF(ADRAT.LT.1.E-12) ADRAT=1.69351
   14 ADVAR=AMIN1(ADRAT,ADVAR)
      IF(ADVAR.GE.1.0) GO TO 8
      ADEXP=0.20
      JADIC=JADIC+1
      IF(JADIC.EQ.1) H1=HINTV
      IF(JADIC.EQ.1) EST1=ESTMAX
      IF(JADIC.EQ.2) GO TO 9
      GO TO 10
    9 ADEXP=ALOG(H1/HINTV)/ALOG(EST1/ESTMAX)
      JADIC=0
   10 CONTINUE
      HINTV=0.9*HINTV*ADVAR**ADEXP
      GO TO 7
    8 CONTINUE
      HINTV=0.9*HINTV*ADVAR**0.20
   13 CONTINUE
      DO 6 I=1,IDIM
    6 Q(I)=Y(I)
      T=TY
      INDIC=0
      IF(TAB.LT.-0.5.OR.FLAG.GT.0.5) RETURN
      GO TO 12
      END
      SUBROUTINE F(NEQN,YP,Y,T,HINTV,INDIC)
C*********************************************************************
C     THIS SUBROUTINE COMPUTES THE RIGHT-HAND SIDES OF THE DIFFERENTIAL
C     EQUATIONS.
C*********************************************************************
      DIMENSION Y(NEQN),YP(NEQN)
      COMMON/ICONST/NQ,NS,N,MODEL,ICBEL
      COMMON/CONST/A1(2),A2(2),A3(2),A7(2),ALFAB(10),ALSO(10),ALB(10),
     *B1(2),B2(2),C1(2),C2(2),C3(2),C4,C5,C6,CAPPAB,CF1(10),CF2(10),
     *ELB(10),EXSIG,FRE(10),FSEFB(10),IFF(10),PI,QQO(2),SIGMA,VN,VT,
     *XIHAT,A6D,ALBEO(10),EM(2),C,STAN,CNS,CBAR,A2D,A3D,DEL,ENCO,EXCB,
     *EDTS(2),ALBE(10),Q(10)
      COMMON/ELCOMP/T1,T2,T3,T4,SLOPE1,SLOPE2,SLOPE,EL
      COMMON/TIME/TSTART,TFIN,HSTEP
C
      IUSE=2
C
      GO TO (1,2,3,4),MODEL
C
C
C     COMPUTATIONS FOR MODELS 1 AND 4
    1 CONTINUE
    4 CONTINUE
      DO 5 I=1,N
C
C     FOR A FAST FIBRE J=1 ELSE J=2
      J=1
      IF(IFF(I).EQ.0)J=2
```

```
      I5=I*5
      I4=I5-1
      I3=I4-1
      I2=I3-1
      I1=I2-1
C
C     COMPUTATION OF STIMULATION FREQUENCIES
      CALL FREQU(T,I,DTI)
      IF(FRE(I).EQ.0.0)GO TO 9
C
C     COMPUTATION OF ALPHA(DTI), WHERE THE TIME DTI OF ACTIVATION IS DEFINED
C     IN SUBROUTINE FREQU.
      CALL ALFT(DTI,PI,FRE(I),ALFA)
      GO TO 10
    9 ALFA=0.0
C
C     COMPUTE THE RIGHT HAND SIDE OF THE DIFFERENTIAL EQUATIONS
C     NOTE THAT BETA=Y(I1) AND GAMMA=Y(I3)
C
   10 YP(I1)=Y(I2)
      YP(I2)=C6*VN*ALFA-C4*Y(I2)-C5*Y(I1)
      YP(I3)=Y(I4)
      YP(I4)=C3(J)*VT*Y(I1)-(C1(J)*Y(I4)+C2(J)*Y(I3))/(CNS/((XIHAT/Y(I5)
     *        )**NS-1.))
      CALL XDTCOM(Y(I5),Y(I3),T,TSTART,NQ,XIHAT,NS,QQO(J),ALBE(I),EXSIG,
     *CF1(I),CF2(I),ALB(I),ALSO(I),A6D,B1(J),B2(J),FSEFB(I),A1(J),
     *A2(J),A3(J),A7(J),YP(I5),HINTV,INDIC,STAN,MODEL,CNS,EDTS(J),Q(I),
     *IUSE,ICBEL)
    5 CONTINUE
      RETURN
C
C
C     COMPUTATIONS FOR MODEL 2
    2 CONTINUE
      DO 6 I=1,N
C
C     FOR A FAST FIBRE J=1 ELSE J=2
      J=1
      IF(IFF(I).EQ.0)J=2
      I2=I*2
      I1=I2-1
C
C     COMPUTATION OF STIMULATION FREQUENCIES
      CALL FREQU(T,I,DTI)
C
      V=.112*FRE(I)/EM(J)
      YP(I1)=EM(J)*(C*V-Y(I1))
      CALL XDTCOM(Y(I2),Y(I1),T,TSTART,NQ,XIHAT,NS,QQO(J),ALBE(I),EXSIG,
     *CF1(I),CF2(I),ALB(I),ALSO(I),A6D,B1(J),B2(J),FSEFB(I),A1(J),
     *A2(J),A3(J),A7(J),YP(I2),HINTV,INDIC,STAN,MODEL,CNS,EDTS(J),Q(I),
     *IUSE,ICBEL)
    6 CONTINUE
      RETURN
C
C
C     COMPUTATIONS FOR MODEL 3
    3 CONTINUE
C     NOTE THAT FSEFB(1)=FSEFB(XIN), FSEFB(2)=FSEFB(XIR), AND FSEFB(3)=FSEFB(X0)
C
C     COMPUTE CONTROLS V(T) AND Z(T)
      CALL CONTRL(T,TSTART,V,Z)
C
      DO 11 L=1,2
      IF(Y(L).LT.0.0.OR.Y(L).GT.1.0) Z=0.0
      IF(Y(L).LT.0.0) Y(L)=0.0
   11 IF(Y(L).GT.1.0) Y(L)=1.0
      EMN=FEMN(CBAR,Y(1),A2D,A3D,DEL)
      EMNR=FEMNR(CBAR,Y(2),Y(1),A2D,A3D,DEL)
      YP(1)=ENDOT(Z,ENCO)
      WP=0.0
      WN=0.0
```

```
      IF(Z.GT.0.0) WP=1.0
      IF(Z.LT.0.0) WN=-1.0
      IF(Z.GT.0.0.AND.Y(1).LT.0.005) Y(3)=Y(4)
      YP(3)=PSIDOT(V,Z,WP,WN,Y(1),Y(3),Y(4),Y(5),DEL,EMN,EMNR,ENCO,XIHAT
     *               ,NS,CBAR,CNS)
      CALL EPSILN(CBAR,Y(1),Y(2),EXCB,NQ,CNS,XIHAT,NS,QQO(1),Y(3),Y(4),
     *            Q(1),Q(2),Y(5),EPSN,EPSR,EPSO,EPSI)
      A7SN=1./FESNR(A2D,A3D,Y(1),Y(2),EPSN,EPSR,EPSO,EPSI)
      CALL XDTCOM(Y(5),EPSI,T,TSTART,NQ,XIHAT,NS,QQO(1),ALBE(1),EXSIG,
     *             CF1(1),CF2(1),ALB(1),ALSO(1),A6D,B1(1),B2(1),FSEFB(1),
     *             A1(1),A2(1),A3(1),A7SN,YP(5),HINTV,INDIC,STAN,MODEL,
     *             CNS,EDTS(1),EPS,IUSE,ICBEL)
      IF(Z.GE.0.0.AND.Y(2).LT.1.E-8) GO TO 7
      YP(2)=ERDOT(Z,WN,Y(2),Y(4),DEL,EMNR,ENCO)
      IF(Z.LT.0.0.AND.Y(2).LT.0.005) Y(4)=Y(3)
      YP(4)=PHIDOT(V,Z,WN,Y(2),Y(3),Y(4),Y(5),DEL,EMN,EMNR,ENCO,XIHAT,
     *              NS,CBAR,CNS)
      GO TO 8
    7 YP(2)=0.0
      YP(4)=0.0
    8 CONTINUE
      RETURN
      END
      SUBROUTINE XDTCOM(XI,GAMMA,T,TSTART,NQ,XIHAT,NS,QQO,ALBE,EXSIG,
     *CF1,CF2,ALB,ALSO,A6D,B1,B2,FSEFB,A1,A2,A3,A7,XIDOT,HINTV,INDIC,
     *STAN,MODEL,CNS,EDTS,Q,IUSE,ICBEL)
C**********************************************************************
C     THIS SUBROUTINE COMPUTES THE RIGHT-HAND SIDES OF THE XIDOT EQUATIONS,
C     AND THE FORCE OUTPUTS FSEFB, IF USED SOLELY FOR THE PURPOSE OF COMPUTING
C     FSEFB, PUT IUSE=0 . IF FSEFB IS KNOWN AND ONLY XIDOT MUST BE COMPUTED,
C     PUT IUSE=1. IF BOTH FSEFB AND XIDOT MUST BE COMPUTED PUT IUSE=2 .
C**********************************************************************
      COMMON/ELCOMP/T1,T2,T3,T4,SLOPE1,SLOPE2,SLOPE,EL
      COMMON/QCOMN/Q1
C     DEFINE INTERNAL FUNCTIONS
      AK1(XY,BE)=2.-EXP(A6D*(XY+BE/ALB-1.))
      AFU(ARG,A2)=-ALOG(ARG)/A2
      ASH(A7,A2,A3,ARG)=A7*(ALOG(AFU(ARG,A2)+SQRT(1.+AFU(ARG,A2)**2))/
     *                 A3-0.5)
C
C     COMPUTATION OF FUNCTIONAL VALUES.
      IF(MODEL.NE.3)AKY5=AMAX1(0.32+0.71*EXP(-1.112*(XI-1.))*SIN(3.722*(
     *                 XI-0.656)),0.005)
      IF(MODEL.EQ.3) AKY5=EXP(-((XI-1.)/STAN)**2)
      TEMP=66200.
      IF(NQ.EQ.3)TEMP=52700.
      RHOO=TEMP
      IF(MODEL.EQ.2)RHOO=TEMP*CNS/((XIHAT/XI)**NS-1.)
      IF(MODEL.NE.3)Q=(QQO+(RHOO*GAMMA)**NQ)/(1.+(RHOO*GAMMA)**NQ)
      IF(MODEL.EQ.3)Q=GAMMA
      IF(T.LT.(TSTART+1.E-10).AND.INDIC.EQ.0)Q1=Q
      AKQ=AKY5*Q
      IF(IUSE.EQ.1) GO TO 7
C
C     COMPUTE INSTANTANEOUS MUSCLE LENGTH.
      CALL LENGTH(T,T1,T2,T3,T4,SLOPE1,SLOPE2,SLOPE,EL0,EL1,EL)
C
      IF(ICBEL.NE.0) GO TO 9
      ALBE=0.0
      GO TO 4
C
C     FIND CROSS-BRIDGE EXTENSIONS ALBE USING NEWTON'S METHOD.
C     SET INITIAL VALUE FOR ALBE.
    9 IF(T.LT.(TSTART+1.E-10))ALBE=ALOG(1.+EXSIG/(1.-QQO))/CF2
    1 F1=EXP(CF1*(EL-ALB*XI-ALBE-ALSO))-1.
      IF(F1.LT.0.)GO TO 4
      F2=AKQ*(EXP(CF2*ALBE)-1.)
      F3=-EXSIG*B1*AK1(XI,ALBE)
      FF=F1-F2-F3
      FDASH=-CF1*(F1+1.)-CF2*(F2+AKQ)-A6D*(2.*B1*EXSIG*F3)/ALB
      IF(ABS(FF).LT.1.E-20)GO TO 4
      ANEWL=ALBE-FF/FDASH
```

```
      RELER=ABS(ANEWL-ALBE)/ABS(ANEWL)
      ALBE=ANEWL
      IF(RELER.GT.0.001)GO TO 1
C
C     END OF NEWTON'S METHOD
C
    4 FSEFB=(EXP(CF1*(EL-ALB*XI-ALBE-ALSO))-1.)/EXSIG
      IF(FSEFB.LT.0.) FSEFB=0.
      IF(IUSE.LT.1)RETURN
C
    7 CONTINUE
      AK1X=AK1(XI,ALBE)
      FCHFB=0.999*(AKQ/(B2*A1)-B1*AK1X)
      IF(FCHFB.LT.0.) FCHFB=0.
      IF(FSEFB.GT.FCHFB) GO TO 8
      FBAK=FSEFB+B1*AK1X
      IF(FBAK.LT.5.E-5)GO TO 3
      TMPORY=AKQ/(B2*FBAK)-A1
      IF(TMPORY.LT.0.0) GO TO 5
      XIDOT=ASH(A7,A2,A3,TMPORY)
      IF(INDIC.EQ.0)Q1=Q
      RETURN
C     THE FOLLOWING PROCEDURE RETURNS THE RELAXING MUSCLE TO ITS EQUILIBRIUM
C     POSITION OR REINITIATES CONTRACTION IF Q INCREASES AGAIN.
    3 XIDOT=A7*0.1*Q
      IF(Q.LT.(QQO+0.001))XIDOT=0.0
      IF(Q.GT.Q1)XIDOT=-A7*Q
      IF(INDIC.EQ.0)Q1=Q
      RETURN
    5 IF(XI.LT.1.0)XIDOT=AMAX1((1.-XI)/HINTV,0.01/HINTV)
      IF(XI.GE.1.0)XIDOT=-0.01/HINTV
      IF(INDIC.EQ.0)Q1=Q
      RETURN
C
C     CHANGE TO LINEAR FORCE-VELOCITY MODEL
    8 FOV=0.01*(1./(A1*B2)-B1-1.)
      XIDOT=A7*(EDTS+((FSEFB+B1*AK1X)/AKQ-0.999/(A1*B2)-0.001*B1)/FOV)
C     ADJUST THE TIME STEP SIZE IF NECESSARY.
      IF(INDIC.EQ.0) HINTV=AMIN1(HINTV,0.01/XIDOT)
      IF(INDIC.EQ.0)Q1=Q
      RETURN
      END
      SUBROUTINE FCXIX(X,F)
C*************************************************************
C     THIS SUBROUTINE COMPUTES THE FUNCTION WHOSE ZERO IS SOUGHT.
C*************************************************************
      COMMON/ICONST/NQ,NS,N,MODEL,ICBEL
      COMMON/CONST/A1(2),A2(2),A3(2),A7(2),ALFAB(10),ALSO(10),ALB(10),
     *B1(2),B2(2),C1(2),C2(2),C3(2),C4,C5,C6,CAPPAB,CF1(10),CF2(10),
     *ELB(10),EXSIG,FRE(10),FSEFB(10),IFF(10),PI,QQO(2),SIGMA,VN,VT,
     *XIHAT,A6D,ALBEO(10),EM(2),C,STAN,CNS,CBAR,A2D,A3D,DEL,ENCO,EXCB,
     *EDTS(2),ALBE(10),Q(10)
      COMMON/ELCOMP/T1,T2,T3,T4,SLOPE1,SLOPE2,SLOPE,EL
      COMMON/INTEGR/IFBR,IFAST
C
      FA=(EXP(CF1(IFBR)*(EL-ALB(IFBR)*X-ALBEO(IFBR)-ALSO(IFBR)))-1.)/
     *   EXSIG
      FSEFB(IFBR)=FA
      IF(FA.LT.0.0)FSEFB(IFBR)=0.0
      IF(MODEL.NE.3)TLX=AMAX1(0.32+0.71*EXP(-1.112*(X-1.))*SIN(3.722*(X-
     *                0.656)),0.005)
      IF(MODEL.EQ.3)TLX=EXP(-((X-1.)/STAN)**2)
      AK1=2.-EXP(A6D*(X+ALBEO(IFBR)/ALB(IFBR)-1.0))
      FB=B1(IFAST)*(TLX-AK1)
      F=FA-FB
      RETURN
      END
      SUBROUTINE FZERO(FCXIX,XIA,XIB,F1,XIX,MAXIT,MES)
C*************************************************************
C     THIS SUBROUTINE USES THE SECANT METHOD TO FIND A ZERO, XIX, OF THE
C     FUNCTION FCXIX.
C*************************************************************
```

```
      ICOUNT=0
      MES=0
      XTEST=0.
1     ICOUNT=ICOUNT+1
      IF(ICOUNT.EQ.1)GO TO 4
      GO TO 5
4     XIO=XIA
      XI1=XIB
      CALL FCXIX(XIO,F10)
      CALL FCXIX(XI1,F11)
5     XI2=XIO-F10*(XI1-XIO)/(F11-F10)
      CALL FCXIX(XI2,F12)
      RELER=ABS(XI2-XTEST)/ABS(XI2)
      IF(RELER.LT.1.E-6)GO TO 3
      IF(ICOUNT.GE.MAXIT)GO TO 2
      XTEST=XI2
      FTEST=F12*F10
      IF(FTEST.GT.0.)GO TO 6
      XI1=XI2
      F11=F12
      GO TO 1
6     XIO=XI2
      F10=F12
      GO TO 1
2     MES=1
3     XIX=XI2
      F1=F12
      RETURN
      END
      FUNCTION ENDOT(Z,ENCO)
C************************************************************************
      ENDOT=ENCO*Z
      RETURN
      END
      FUNCTION ERDOT(Z,WN,YR,YPH,DEL,EMNR,ENCO)
C************************************************************************
      ERDOT=-ENCO*Z*(YR-WN*DEL)/(YR*DEL)-(1.+WN)*EMNR*YR/(1.E-3*EMNR+
     *      (YPH/1.373E-8)**2)
      RETURN
      END
      FUNCTION PSIDOT(V,Z,WP,WN,YN,YPS,YPH,YXN,DEL,EMN,EMNR,ENCO,XIMIN,
     *                NS,CBAR,CNS)
C************************************************************************
      ROXON=53300.*CNS/((XIMIN/YXN)**NS-1.)
      PSIDOT=EMN*(1.373E-4*V-YPS)+WP*Z*CBAR*ENCO*(1.-EXP(ROXON*(YPS-
     *       YPH)))/(ROXON*(1.-EXP(-CBAR*YN*DEL)))-(1.+WN)*YPH*EMNR
      RETURN
      END
      FUNCTION PHIDOT(V,Z,WN,YR,YPS,YPH,YXR,DEL,EMN,EMNR,ENCO,XIMIN,NS,
     *                CBAR,CNS)
C************************************************************************
      ROXOR=53300.*CNS/((XIMIN/YXR)**NS-1.)
      PHIDOT=-EMNR*YPH-WN*(EMN*YPH*(1.373E-4*V/(YPS*DEL)-1.)-Z*CBAR*
     *       ENCO*(1.-EXP(ROXOR*(YPH-YPS)))/(ROXOR*(EXP(CBAR*YR*DEL)-
     *       1.)))
      RETURN
      END
      FUNCTION FEMN(CBAR,EN,A2D,A3D,DEL)
C************************************************************************
      E1(X)=(X**2+2.334733*X+.250621)/(X**2+3.330657*X+1.681534)
      AA2=A2D/0.372
      AA3=A3D/0.372
      EX=EN+DEL
      X1=CBAR*(A2D/A3D-EX)
      X2=CBAR*A2D/A3D
      FEMN=(EXP(CBAR*EX)*E1(X1)/(AA2-AA3*EX)-E1(X2)/AA2)/(EXP(CBAR*EX)
     *     -1.0)
      RETURN
      END
      FUNCTION FEMNR(CBAR,R,EN,A2D,A3D,DEL)
C************************************************************************
      E1(X)=(X**2+2.334733*X+.250621)/(X**2+3.330657*X+1.681534)
      AA2=A2D/0.372
      AA3=A3D/0.372
      EX=EN+DEL
      U=R+DEL
      X1=CBAR*(A2D/A3D-EX-U)
      X2=CBAR*(A2D/A3D-EX)
      FEMNR=(E1(X1)/(AA2-AA3*(EX+U))-E1(X2)*EXP(-CBAR*U)/(AA2-AA3*EX))
```

195

```
            /(1.-EXP(-CBAR*U))
      RETURN
      END
      FUNCTION FESNR(A2D,A3D,EN,R,EPSN,EPSR,EPSO,EPSI)
C***************************************************************************
      SFUN(A,B,A2D,A3D)=(A2D-A3D*(A+B*(0.568+0.2307*B)))/0.297
      ENR=EN+R
      ANR=1.0-ENR


      FESNR=(EPSN*SFUN(0.0,EN,A2D,A3D)+EPSR*SFUN(EN,R,A2D,A3D)+EPSO*
     *       SFUN(ENR,ANR,A2D,A3D))/EPSI
      RETURN
      END
      SUBROUTINE EPSILN(CBAR,EN,R,EXCB,NQ,CNS,XIHAT,NS,QQO,PSI,PHI,QPSI,
     *                 QPHI,XI,EPSN,EPSR,EPSO,EPSI)
C***************************************************************************
      TEMP=66200.0
      IF(NQ.EQ.3)TEMP=52700.0
      RHOO=TEMP*CNS/((XIHAT/XI)**NS-1.0)
      QPSI=(QQO+(RHOO*PSI)**NQ)/(1.+(RHOO*PSI)**NQ)
      QPHI=(QQO+(RHOO*PHI)**NQ)/(1.+(RHOO*PHI)**NQ)
      EXNR=EXP(CBAR*(EN+R))
      EXN=EXP(CBAR*EN)
      EPSN=(EXN-1.)*QPSI/EXCB
      EPSR=(EXNR-EXN)*QPHI/EXCB
      EPSO=(EXCB+1.-EXNR)*QQO/EXCB
      EPSI=EPSN+EPSR+EPSO
      RETURN
      END
      SUBROUTINE PLOTR(OUTINT,ELIN)
C***************************************************************************
C     THIS ROUTINE INITIATES THE PLOTTING PROCEDURE.
C***************************************************************************
      DIMENSION YY(1550,8),X(1550),Y(8)
      COMMON/ICONST/NQ,NS,N,MODEL,ICBEL
      COMMON/TIME/TSTART,TFIN,HSTEP
      COMMON/CGRAPH/IGRAPH,IFIB,NPLOTS,INDEX(5),IRUN,LINE
      COMMON/CIOF/LR,LW,IOUT1,IOUTP
C     FIRSTX, FIRSTY DENOTE THE X,Y COORDINATES OF THE ORIGIN, WHILE XLENTH,
C     YLENTH DESIGNATE THE ACTUAL LENGTHS OF THE COORDINATE AXES IN INCHES.
      DATA FIRSTX,FIRSTY/0.0,0.0/,XLENTH,YLENTH/10.,10./
C     NP IS THE NUMBER OF POINTS TO BE PLOTTED (MAXIMUM=500 PER PLOT).
      NP=(INT((TFIN-TSTART)/OUTINT+1.E-3)+1)*IRUN
      NP2=NP+2
      YMAX=1.0
C     NORMALIZING FACTORS FOR BETA AND GAMMA:
      BETAN=3.0E-08
      GAMMAN=1.373E-04
C     THE FOLLOWING PARAMETER IS USED TO TERMINATE PLOTTING
      IPEN=999
      NY=7
      IF(MODEL.EQ.2)NY=6
C
C     READ DATA FROM FILE IOUT1.
C
      DO 1 I=1,NP
      READ(IOUT1,10)X(I),(Y(J),J=1,NY)
      DO 1 K=1,NPLOTS
      J=INDEX(K)
C     NORMALIZE BETA (FOR MODELS 1 AND 4 ONLY) , GAMMA
C     (FOR MODELS 1,2 AND 4) AND LENGTH (FOR ALL MODELS).
      IF((MODEL.EQ.1.OR.MODEL.EQ.4).AND.J.EQ.1)Y(J)=Y(J)/BETAN
      IF(((MODEL.EQ.1.OR.MODEL.EQ.4).AND.J.EQ.2).OR.(MODEL.EQ.2.AND.J.
     *EQ.1))Y(J)=Y(J)/GAMMAN
      IF(J.EQ.NY)Y(J)=Y(J)/ELIN
      YMAX=AMAX1(YMAX,Y(J))
    1 YY(I,J)=Y(J)
      IF(YMAX.GT.1.0)YMAX=2.0
      IF(YMAX.GT.1.0)GO TO 4
      YMAX=1.1
      YLENTH=11.0
C
C     ADD THE SCALING PARAMETERS TO THE END OF THE VARIABLE VECTORS.
C
```

```
    4 CALL VSCALE(X,NP2,XLENTH,FIRSTX,TFIN)
      DO 3 L=1,NPLOTS
      J=INDEX(L)
    3 CALL VSCALE(YY(1,J),NP2,YLENTH,FIRSTY,YMAX)
C
C     DRAW COORDINATE SYSTEM.
C
      J=INDEX(NPLOTS)
      CALL AXES(X,YY(1,J),NP2,XLENTH,YLENTH)
C
C     DRAW LINES BETWEEN THE DATA POINTS.
C
      DO 2 K=1,NPLOTS
      J=INDEX(K)
    2 CALL GRAPH(X,YY(1,J),NP2)
C
C     TERMINATE THE PLOTTING
C
      CALL PLOT(FIRSTX,FIRSTY,IPEN)
   10 FORMAT(9E13.6)
      RETURN
      END
      SUBROUTINE VSCALE(V,N,ALENTH,FIRSTV,VMAX)
C***********************************************************************
C     THIS ROUTINE ADDS THE TWO ELEMENTS FIRSTV AND DELTAV TO THE VECTOR V.
C     ALENTH = LENGTH OF AXIS IN INCHES.
C     FIRSTV = ORIGIN ON AXIS.
C     DELTAV = SCALE INCREMENT ON AXIS.
C     VMAX   = MAXIMUM VALUE OF ALL THE ELEMENTS IN V.
C***********************************************************************
      DIMENSION V(N)
      DELTAV=VMAX/ALENTH
      NM1=N-1
      V(NM1)=FIRSTV
      V(N)=DELTAV
      RETURN
      END
      SUBROUTINE AXES(X,Y,N,XL,YL)
C***********************************************************************
C     THIS ROUTINE DRAWS THE X AND Y-AXIS AND WRITES THE CORRESPONDING CAPTIONS.
C     XL = REAL LENGTH OF THE X-AXIS IN INCHES.
C     YL = REAL LENGTH OF THE Y-AXIS IN INCHES.
C***********************************************************************
      DIMENSION X(N),Y(N),NINCH(2)
      COMMON/CIOF/LR,LW,IOUT1,IOUTP
      DATA NINCH(1),NINCH(2),XINCH,YINCH,IPEN/18,18,0.8,0.8,-3/,
     *     XPAGE,YPAGE,XDEG,YDEG/0.0,0.0,0.0,90.0/
      CALL PLOTS(NINCH,6HMYOSIM,IOUTP)
C
C     SET PEN TO ORIGIN AND DRAW AXES USING CALCOMP ROUTINES.
C
      NM1=N-1
      CALL PLOT(XINCH,YINCH,IPEN)
      CALL AXIS(XPAGE,YPAGE,15H                ,-15,XL,XDEG,X(NM1),X(N))
      CALL AXIS(XPAGE,YPAGE,26H                          ,26,YL,YDEG,
     *          Y(NM1),Y(N))
      CALL SYMBOL(3.,-.7,.21,15HTIME IN SECONDS,0.,15)
      CALL SYMBOL(-.45,3.,.21,25HNORMALIZED MODEL RESPONSE,90.0,25)
      RETURN
      END
      SUBROUTINE GRAPH(X,Y,N)
C***********************************************************************
C     THIS SUBROUTINE DRAWS LINES BETWEEN THE N-2 POINTS (X,Y).
C***********************************************************************
      DIMENSION X(N),Y(N)
      COMMON/TIME/TSTART,TFIN,HSTEP
      COMMON/CGRAPH/IGRAPH,IFIB,NPLOTS,INDEX(5),IRUN,LINE
      DATA FIRSTX,FIRSTY,IPEN2,IPEN3/0.0,0.0,2,3/
      M=N-2
C
C     DRAW A LINE THROUGH THE M POINTS.
C
      DO 1 I=1,M
C     SCALE THE DATA TO FIT ONTO THE AXES
```

```
          XX=X(I)/X(N)
          Y(I)=Y(I)/Y(N)
   C
          IF(ABS(XX-TSTART).LT.1.E-8)CALL PLOT(FIRSTX,FIRSTY,IPEN3)
          IPEN=IPEN2
          IF(LINE.EQ.1)GO TO 2
          IF(I.GT.M/IRUN.AND.MOD(I,2).EQ.1)IPEN=IPEN3
        2 CONTINUE
          CALL PLOT(XX,Y(I),IPEN)
        1 CONTINUE
          RETURN
          END
          SUBROUTINE ALFT(T,PI,FRE,ALFA)
   C***************************************************************
   C     THIS SUBROUTINE GENERATES POSITIVE 1 MSEC HALF-SINE IMPULSES ALFA(T),
   C     AT A FREQUENCY OF FRE HERTZ.
   C***************************************************************
          NCOMP=INT(T*FRE)
          TCOMP=FLOAT(NCOMP)/FRE+.001
          ALFA=0.0
          IF(T.LE.TCOMP)ALFA=SIN(1000.*PI*(T-TCOMP+.001))
          RETURN
          END
          SUBROUTINE LENGTH(T,T1,T2,T3,T4,SLOPE1,SLOPE2,SLOPE,ELO,EL1,EL)
   C***************************************************************
   C     THIS SUBROUTINE COMPUTES THE INSTANTANEOUS MUSCLE LENGTH.
   C***************************************************************
   C     EL IS CONSTANT FOR T GREATER THAN TSTART AND T LESS THAN T1.
          SLOPE=0.
          IF(T.GE.T1)GO TO 1
          ELO=EL
          RETURN
   C
        1 IF(T.GE.T2)GO TO 2
          SLOPE=SLOPE1
          EL1=ELO+SLOPE1*(T-T1)
          EL=EL1
          RETURN
   C
   C     EL=EL(T2) FOR T GREATER THAN T2 AND T LESS THAN T3.
        2 IF(T.GE.T3.AND.T.LT.T4)GO TO 3
          RETURN
        3 SLOPE=SLOPE2
          EL=EL1+SLOPE2*(T-T3)
   C     EL=EL(T4) FOR T GREATER THAN T4.
          RETURN
          END
          SUBROUTINE CONTRL(T,TSTART,V,Z)
   C***************************************************************
   C     THIS SUBROUTINE GENERATES THE CONTROL FUNCTIONS V(T) AND Z(T).
   C***************************************************************
          V=0.0
          Z=0.0
          IF(T.GE.TSTART) V=1.0
          IF(T.GE.TSTART) Z=1.0
          IF(T.GE.0.07) Z=0.0
          IF(T.GE.0.50) Z=-1.0
          IF(T.GE.0.57) Z=0.0
          RETURN
          END
          SUBROUTINE FREQU(T,J,DTI)
   C***************************************************************
   C     THIS ROUTINE COMPUTES THE INSTANTANEOUS STIMULATION FREQUENCIES FRE(I),
   C     I=1,...,N (MAXIMUM(N)=10), FOR N MUSCLE FIBRES OR MOTOR UNITS. THE TIME
   C     FUNCTIONS FRE(I,T) MUST BE SPECIFIED BY THE USER. A DELAY OPERATOR
   C     PRODUCING DELAYED ACTIVATION IS ALSO INCLUDED.
   C***************************************************************
          COMMON/ICONST/NQ,NS,N,MODEL,ICBEL
          COMMON/CONST/A1(2),A2(2),A3(2),A7(2),ALFAB(10),ALSO(10),ALB(10),
         *B1(2),B2(2),C1(2),C2(2),C3(2),C4,C5,C6,CAPPAB,CF1(10),CF2(10),
         *ELB(10),EXSIG,FRE(10),FSEFB(10),IFF(10),PI,QQO(2),SIGMA,VN,VT,
         *XIHAT,A6D,ALBEO(10),EM(2),C,STAN,CNS,CBAR,A2D,A3D,DEL,ENCO,EXCB,
```

```
     *EDTS(2),ALBE(10),Q(10)
      COMMON/CGRAPH/IGRAPH,IFIB,NPLOTS,INDEX(5),IRUN,LINE
C
      GO TO (1,2,3,4),MODEL
    1 CONTINUE
      DTI=T
C     INSERT FUNCTION DEFINITION FRE(I,T) FOR MODEL 1 HERE.
      RETURN
    2 CONTINUE
      DTI=T
C     INSERT FUNCTION DEFINITION FRE(I,T) FOR MODEL 2 HERE.
    3 RETURN
    4 CONTINUE
      DTI=T
C     INSERT FUNCTION DEFINITION FRE(I,T) FOR MODEL 4 HERE.
C
C     DELAY OPERATOR
      F=FRE(J)
      DTI=T-FLOAT(J-1)/(F*FLOAT(N))
      IF(DTI.LT.0.0)DTI=0.0
      RETURN
      END
```

# Bibliography

ABBOTT, B.C. The heat production associated with the maintenance of a prolonged contraction and the extra heat produced during large shortening. J. Physiol. *112* (1951), 438–445.

ABRAMOWITZ, M., STEGUN, I.A. Handbook of mathematical functions. Dover Publ., New York, 1968.

ADRIAN, R.H., CHANDLER, W.K., HODGKIN, A.L. The kinetics of mechanical activation in frog muscle. J. Physiol. *204* (1969), 207–230.

ALEXANDER, R.S., JOHNSON, P.D. Muscle stretch and theories of contraction. Am. J. Physiol. *208* (1965), 412–416.

ASTBURY, W.T. X-ray studies of muscle. Proc. Roy. Soc. B, *137* (1950), 58.

BAGUST, J., KNOTT, S., LEWIS, D.M. Isometric contractions of motor units in a fast twitch muscle of the cat. J. Physiol. *231* (1973), 87–104.

BAHLER, A.S. Series elastic component of mammalian skeletal muscle. Am. J. Physiol. *213* (1967), 1560–1564.

BAHLER, A.S., FALES, J.T., ZIERLER, K.L. The active state of mammalian skeletal muscle. J. Gen. Physiol. *50* (1967), 2239–2253.

BAHLER, A.S. Modeling of mammalian skeletal muscle. IEEE Trans. BioMed. Eng. *BME-15* (1968), 249–257.

BAHLER, A.S., FALES, J.T., ZIERLER, K.L. The dynamic properties of mammalian skeletal muscle. J. Gen. Physiol. *51* (1968), 369–384.

BALLREICH, R. An analysis of long-jump. In: Biomechanics III, pp. 394–402, Medicine and Sport, vol. 8. Karger, Basel, 1973.

BANUS, M.G., ZETLIN, A.M. The relation of isometric tension to length in muscle. J. cell. comp. Physiol. *12* (1938), 403–420.

BÁRÁNY, M. ATPase activity of myosin correlated with speed of muscle shortening. J. Gen. Physiol. *50* (1967), 197–218.

BAWA, P., MANNARD, A., STEIN, R.B. Predictions and experimental tests of a visco-elastic muscle model using elastic and inertial loads. Biol. Cybernetics *22* (1976), 139–145.

BENDALL, J.R. Muscles, molecules and movement. Heinemann Educ. Books Ltd., London, 1969.

BERNSTEIN, J. Experimentelles und eine kritische Theorie der Muskelkontraktion. Arch. f. ges. Physiol. *162* (1915), 1.

BIGLAND, B., LIPPOLD, O.C.J. Motor unit activity in the voluntary contraction of human muscle. J. Physiol. *125* (1954), 322–335.

BOLSTAD, G., ERSLAND, A. Energy metabolism in different human skeletal muscles during voluntary contraction. Acta Physiol. Scand. *95* (1975), 73A–74A.

BRANDELL, B.R. An analysis of muscle coordination in walking and running gaits. In: Biomechanics III, pp. 278–287, Medicine and Sport, vol. 8. Karger, Basel, 1973.

BRETT, J.R. The swimming energetics of salmon. Scient. Amer. *213*, No. 2 (1965), 76–84.

BUCHTHAL, F., KAISER, E. The rheology of the cross striated muscle fibre with particular reference to isotonic conditions. Dan. Biol. Medd. *21* (1951), 318.

BULLER, A.J., ECCLES, J.C., ECCLES, R.M. Differentiation of fast and slow muscles in the cat hind limb. J. Physiol. *150* (1960), 399–416.

BULLER, A.J., LEWIS, D.M. The rate of tension development in isometric tetanic contractions of mammalian fast and slow skeletal muscle. J. Physiol. *176* (1965), 337–354.

BURKE, R.E., LEVINE, D.N., ZAJAC, F.E., TSAIRIS, P., ENGEL, W.K. Mammalian motor units: physiological-histochemical correlation in three types in cat gastrocnemius. Science *174* (1971), 709–712.

CHAO, E.Y., RIM, K. Application of optimization principles in determining the applied moments in human leg joints during gait. J. Biomechanics *6* (1973), 497–510.

CHOW, C.K., JACOBSON, D.H. Studies of human locomotion via optimal programming. Math. Biosci. *10* (1971), 239–306.

CLOSE, R.I. Dynamic properties of fast and slow skeletal muscles of the rat during development. J. Physiol. *173* (1964), 74–95.

CLOSE, R.I. The relation between intrinsic speed of shortening and duration of the active state of muscle. J. Physiol. *180* (1965), 542–559.

CLOSE, R.I. Dynamic properties of fast and slow skeletal muscles of the rat after nerve cross-union. J. Physiol. *204* (1969), 331–346.

CLOSE, R.I. Dynamic properties of mammalian skeletal muscle. Physiol. Rev. *52* (1972), 129–197.

CODDINGTON, E.A., LEVINSON, N. Theory of ordinary differential equations (pp. 13–61). McGraw-Hill, New York, 1955.

CONSTANTIN, L.L., PODOLSKY, R.J. Evidence for depolarization of the internal membrane system in activation of frog semitendinosus muscle. Nature *210* (1966), 483–486.

COTES, J.E., MEADE, F. Energy expenditure and energy demand in walking. Ergonomics *3* (1960).

CROWE, A. A mechanical model of muscle and its application to the intrafusal fibres of the mammalian muscle spindle. J. Biomechanics *3* (1970), 583–592.

DAREMBERG, C.V. (transl.). Oeuvres anatomiques, physiologiques et médicales de Galien. Two vols., Paris, 1854–57.

DAVIES, R.E. A molecular theory of muscle contraction: calcium dependent contractions with hydrogen bond formation plus ATP-dependent extensions of part of the myosin-actin cross-bridges. Nature *199* (1963), 1068–1074.

DEL CASTILLO, J., KATZ, B. Biophysical aspects of neuro-muscular transmission. Progr. Biophys. *6* (1956), 121–170.

DÉLÉZE, J.B. The mechanical properties of the semitendinosus muscle at lengths greater than its length in the body. J. Physiol. *158* (1961), 154–164.

DESMEDT, J.E., GODAUX, E. Ballistic contractions in man: characteristic recruitment pattern of single motor units of the tibialis anterior muscle. J. Physiol. *264* (1977), 673–693.

DESMEDT, J.E., HAINAUT, K. Kinetics of myofilament activation in potentiated contraction: Staircase phenomenon in human skeletal muscle. Nature *217* (1968), 529–532.

DRAGOMIR, C.T. On the nature of forces acting between myofilaments in resting state and under contraction. J. theor. Biol. *27* (1970), 343–356.

EBASHI, S., KODAMA, A. A new protein factor promoting aggregation of tropomyosin. J. Biochem. *58* (1965), 107.

EBASHI, S., ENDO, M. Calcium ion and muscular contraction. Progr. Biophys. Mol. Biol. *18* (1968), 125–183.

EBASHI, S., KODAMA, A., EBASHI, F. Troponin. 1. Preparation and physiological function. J. Biochem. *64* (1968), 465.

EBERSTEIN, A., GOODGOLD, J. Slow and fast twitch fibres in human skeletal muscle. Amer. J. Physiol. *215* (1968), 535–541.

ECCLES, J.C. The understanding of the brain; p. 20. McGraw-Hill, New York, 1973.

EDMAN, K.A.P., MULIERI, L.A., SCUBON-MULIERI, B. Non-hyperbolic force-velocity relationship in single muscle fibres. Acta Physiol. Scand. *98* (1976), 143–156.

EDMAN, K.A.P. The velocity of unloaded shortening and its relation to sarcomere length and isometric force in vertebrate muscle fibres J. Physiol. *291* (1979), 143–159.

EDMAN, K.A.P., ELZINGA, G., NOBLE, M.I.M. Enhancement of mechanical performance by stretch during tetanic contractions of vertebrate skeletal muscle fibres. J. Physiol. *218* (1978), 139–155.

EDWARDS, R.H.T., HILL, D.K., JONES, D.A. Heat production and chemical changes during isometric contractions of the human quadriceps muscle. J. Physiol. *251* (1975), 303–315.

ELLIOT, G.F., ROME, E.M., SPENCER, M. A type of contraction hypothesis applicable to all muscles. Nature 226 (1970), 417.

ENGELHARDT, V.A., LYUBIMOVA, M.N. On the mechanochemistry of muscle. Biokhimiya 7 (1942), 205.

ENGELMANN, T.W. Mikroskopische Untersuchungen über die quergestreifte Muskelsubstanz. II. Arch. f. ges. Physiol. 7 (1873), 155.

FALK, G. Predicted delays in the activation of the contractile system. Biophys. J. 8 (1968), 608–625.

FALK, G., FATT, P. Linear electrical properties of striated muscle fibres observed with intracellular electrodes. Proc. Roy. Soc. B 160 (1964), 69–123.

FATT, P. Skeletal neuromuscular transmission. In: Neurophysiology I, p. 204. Am. Physiol. Soc., Washington, 1959.

FARQUHARSON, A.S.L. (transl.). (Aristotle's) De motu animalium. In: The works of Aristotle, tr. ed. Smith & Ross, Vol. 5. Claredon Press, Oxford, 1912.

FISCHER, O. Theoretische Grundlagen für eine Mechanik der lebenden Körper. B.G. Teubner, Berlin, 1906.

FITZHUGH, R. A model of optimal voluntary muscular control. J. Math. Biol. 4 (1977), 203–236.

FLITNEY, F.W., HIRST, D.G. Cross-bridge detachment and sarcomere 'give' during stretch of active frog's muscle. J. Physiol. 276 (1978), 449–465.

FUNG, Y.C.B. Biomechanics: Its scope, history and some problems of continuum mechanics in physiology. Appl. Mech. Rev. 21 (1968), 1–20.

FUNG, Y.C.B. Mathematical representation of the mechanical properties of heart muscle. J. Biomechanics 3 (1970), 381–404.

FUNG, Y.C.B. Chapter 7 of: Biomechanics. Its foundations and objectives. Prentice-Hall, New Jersey, 1972.

GIBBS, C.L., GIBSON, W.R. Energy production of rat soleus muscle. Amer. J. Physiol. 223 (1972), 864–871.

GILLIS, J.M. The site of action of Ca in producing contraction in striated muscle. J. Physiol. 200 (1969), 849–864.

GLANTZ, S.A. A constitutive equation for the passive properties of muscle. J. Biomechanics 7 (1974), 137–145.

GONZALEZ-SERRATOS, H. Inward spread of contraction during a twitch. J. Physiol. 185 (1966), 20–21 P.

GORDON, A.M., HUXLEY, A.F., JULIAN, F.J. The variation in isometric tension with sarcomere length in vertebrate muscle fibres. J. Physiol. 184 (1966), 170–192.

GREEN, D.G. A note on modeling muscle in physiological regulators. Med. & biol. Eng. 7 (1969), 41–48.

GRIMBY, L., HANNERZ, J. Firing rate and recruitment order of toe extensor motor units in different modes of voluntary contraction. J. Physiol. 264 (1977), 864–879.

GRIMBY, L., HANNERZ, J., HEDMAN, B. Contraction time and voluntary discharge properties of individual short toe extensor motor units in man. J. Physiol. *289* (1979), 191–201.

GYDIKOV, A.A., KOSAROV, D.S. Physiological characteristics of the tonic and phasic motor units in human muscles. In: Motor control, A.A. Gydikov, N.T. Tankov, D.S. Kosarov (eds.), Plenum, New York, 1973.

HARDT, D.E. Determining muscle forces in the leg during normal human walking - an application and evaluation of optimization methods. J. Biomech. Eng. Trans. ASME *100* No. 2 (1978), 72–78.

HARTSHORNE, D.J., DREIZEN, P. Studies on the subunit composition of troponin. Cold Spring Harb. Symp. Quant. Biol. XXXVII (1973), 225–234.

HATZE, H. A theory of contraction and a mathematical model of striated muscle. J. theor. Biol. *40* (1973), 219–246.

HATZE, H. A model of skeletal muscle suitable for optimal motion problems. In: Biomechanics IV, pp. 417–422, S. Karger, Basel, 1974.

HATZE, H. A control model of skeletal muscle and its application to a time-optimal bio-motion. Ph.D. Thesis, University of South Africa, Pretoria, 1975a.

HATZE, H. A new method for the simultaneous measurement of the moment of inertia, the damping coefficient and the location of the centre of mass of a body segment in situ. Europ. J. Appl. Physiol. *34* (1975b), 217–226.

HATZE, H. The complete optimization of a human motion. Math. Biosci. *28* (1976), 99–135; with correction in vol. *37* (1977), p. 279.

HATZE, H. A myocybernetic control model of skeletal muscle. Biol. Cybernetics *25* (1977a), 103–119.

HATZE, H. A complete set of control equations for the human musculoskeletal system. J. Biomechanics *10* (1977b), 799–805.

HATZE, H. A general myocybernetic control model of skeletal muscle. Biol. Cybernetics *28* (1978a), 143–157.

HATZE, H. A teleological explanation of Weber's law and the motor unit size law. Bull. Mathem. Biol. *41* (1979), 407–425.

HATZE, H. Estimation of myodynamic parameter values from observations on isometrically contracting muscle groups. CSIR Techn. Rpt. TWISK 153, Pretoria, 1980a.

HATZE, H. Neuromusculoskeletal control systems modelling - a critical survey of recent developments. IEEE Trans. Autom. Control *25* (1980b), No. 3, 375–385.

HATZE, H. A comprehensive model for human motion simulations and its application to the take-off phase of the long jump. J. Biomechanics (1980c, in press).

HATZE, H. A mathematical model for the computational determination of parameter values of anthropomorphic segments. J. Biomechanics *13* (1980d), 833–843.

HATZE, H. The use of optimally regularized Fourier series for estimating higher-order derivatives of noisy biomechanical data. J. Biomechanics 14 (1980e), 13–18.

HATZE, H., BUYS, J.D. Energy-optimal controls in the mammalian neuromuscular system. Biol. Cybernetics 27 (1977), 9–20.

HAUT, R.C., LITTLE, R.W. A constitutive equation for collagen fibres. J. Biomechanics 5 (1972), 423–430.

HAYES, K.C., HATZE, H. Passive visco-elastic properties of the structures spanning the human elbow joint. Europ. J. appl. Physiol. 37 (1977), 265–274.

HEFNER, L.L., BOWEN, T.E. Elastic components of cat papillary muscle. Am. J. Physiol. 212 (1967), 1221–1227.

HEMAMI, H., WEIMER, F.C., KOOZEKANANI, S.H. Some aspects of the inverted pendulum problem for modelling of locomotion systems. IEEE Trans. Automat. Contr. AC–18 (1973), 658–661.

HEMAMI, H., GOLLIDAY, C.L. The inverted pendulum and biped stability. Math. Biosci. 34 (1977), 95–110.

HEMAMI, H., FARNSWORTH, R.L. Postural and gait stability of a planar five link biped by simulation. IEEE Trans. Autom. Contr. AC–22 (1977), 452–458.

HENNEMAN, E., SOMJEN, G., CARPENTER, D.O. Excitability and inhibitability of motoneurones of different sizes. J. Neurophysiol. 28 (1965), 597–620.

HENNEMAN, E. Peripheral mechanisms involved in the control of muscle. In: Mountcastle, V.B. (ed.) Medical Physiology, pp. 1697–1716; Mosby, St. Louis, 1968.

HENNEMAN, E., OLSON, C.B. Relations between structure and function in the design of skeletal muscle. J. Neurophysiol. 28 (1965), 581–598.

HILL, A.V. Muscular exercise. Nature 112 (1923), 77.

HILL, A.V. The heat of shortening and the dynamic constants of muscle. Proc. Roy. Soc. 126 B (1938), 136–195.

HILL, A.V. The abrupt transition from rest to activity in muscle. Proc. Roy. Soc. B136 (1949), 399–420.

HILL, A.V. The series elastic component of muscle. Proc. Roy. Soc. B 137 (1950), 273–280.

HILL, A.V. A discussion on muscular contraction and relaxation: their physical and chemical basis. Proc. Roy. Soc. (Lond.) B 137 (1950b), 40.

HILL, A.V. The instantaneous elasticity of active muscle. Proc. Roy. Soc. B 141 (1953), 161.

HILL, D.K. Tension due to interaction between the sliding filaments in resting striated muscle. The effect of stimulation. J. Physiol. 199 (1968), 637–684.

HILL, L. A-band length, striation spacing and tension change on stretch of active muscle. J. Physiol. 266 (1977), 677–685.

HODGKIN, A.L., HOROWICZ, P. Potassium contractures in single muscle fibres. J. Physiol. *153* (1960), 386–403.
HUXLEY, A.F., TAYLOR, R.E. Local activation of striated muscle fibres. J. Physiol. *144* (1958), 426–441.
HUXLEY, A.F. Muscle structure and theories of contraction. Progr. Biophys. *7* (1957), 255.
HUXLEY, A.F., PEACHEY, L.D. Local activation of crab muscle. J. Cell Biol. *23* (1964), 107a.
HUXLEY, A.F., SIMMONS, R.M. Proposed mechanism of force generation in striated muscle. Nature *233* (1971), 533–538.
HUXLEY, H.E. Electron microscope studies of the organisation of the filaments in striated muscle. Biochem. biophys. Acta *12* (1953), 387.
HUXLEY, H.E. The double array of filaments in cross-striated muscle. J. biophys. biochem. Cyt. *3* (1957), 631.
HUXLEY, H.E. The mechanism of muscular contraction. Science *164* (1969), 1356–1366.
INESI, G. Active transport of calcium ion in sarcoplasmic membranes. Ann. Rev. Biophys. Bioeng. *1* (1972), 191–210.
JEWELL, B.R., WILKIE, D.R. An analysis of the mechanical components in frog's striated muscle. J. Physiol. *143* (1958), 515–540.
JEWELL, B.R., WILKIE, D.R. The mechanical properties of relaxing muscle. J. Physiol. *152* (1960), 30–47.
JÖBSIS, F.F., O'CONNOR, M.J. Calcium release and reabsorption in the sartorius muscle of the toad. Biochem. Biophys. Res. Com. *25* (1966), 246–252.
JOYCE, G.C., RACK, P.M.H. Isotonic lengthening and shortening movements of cat soleus muscle. J. Physiol. *204* (1969), 475–491.
JOYCE, G.C., RACK, P.M.H., WESTBURY, D.R. The mechanical properties of cat soleus muscle during controlled lengthening and shortening movements. J. Physiol. 204(1969), 461–474.
JULIEN, F.J. The effect of calcium on the force-velocity relation of briefly glycerinated frog muscle fibres. J. Physiol. *218* (1971), 117.
KELLY, D.L. Kinesiology. Prentice-Hall, New Jersey, 1971.
KOMI, P.V. Measurement of the force-velocity relationship in human muscle under concentric and eccentric contractions. In: Biomechanics III (Cerquiglini, A., Venerando, A., and Wartenweiler, J., ed.), pp. 224–229, S. Karger, Basel, 1973.
KÜHN, K.G. (transl.) (Galen's) Opera omnia. Latin with orig. greek text, 20 Vols. Leipzig, 1821–33.
KUFFLER, S.W., NICHOLLS, J.G. From neuron to brain, pp. 183–188. Sinauer Assoc., Sunderland (Mas.), 1976.
LAMMERT, O., JØRGENSEN, F., EINER-JENSEN, N. Accelerometermyography (AMG) I: method for measuring mechanical vibrations from isometrically contracted muscles. In: Biomechanics V-A (Komi, P.V., ed.), pp. 152–158. University Park Press, Baltimore, 1976.

LITTRÉ, E. (transl.). On the nature of bones. Vol. IX of Oeuvres complétes d'Hippocrate. Paris, 1839–61.

LOEWY, A.G. A theory of covalent bonding in muscle contraction. J. theor. Biol. *20* (1968), 164.

LOWEY, S., SLAYTER, H.S., WEEDS, A.G., BAKER, H. Substructure of the myosin molecule. J. Mol. Biol. *42* (1969), 1–29.

LUHTANEN, P., KOMI, P.V. Mechanical power and segmental contribution to force impulses in long jump take-off. Europ. J. appl. Physiol. *41* (1979), 267–274.

MARSDEN, C.D., MEADOWS, J.C., MERTON, P.A. Isolated single motor units in human muscle and their rate of discharge during maximal voluntary effort. J. Physiol. *217* (1971), 12–13 P.

McCROREY, H.L., GALE, H.H., ALPERT, N.R. Mechanical properties of cat tenuissimus muscle. Am. J. Physiol. *210* (1966), 114–120.

McDOUGALL, W. On the structure of cross-striated muscle, and a suggestion as to the nature of its contraction. J. Anat. and Physiol. *31* (1897), 410.

MEIGS, E.B. The structure of the element of cross-striated muscle, and the changes of form which it undergoes during contraction. Z. f. allg. Physiol. *8* (1908), 81.

MENDELL, L.M., HENNEMAN, E. Terminals of single Ia fibres: location, density and distribution within a pool of 300 homonymous motoneurons. J. Neurophysiol. *34* (1971), 171–187.

MEYER, K.H. Uber Feinbau, Festigkeit und Kontraktilität tierischer Gewebe. Biochem. Z. *214* (1929), 253.

MEYERHOF, O. Über einige Probleme der Muskelphysiologie. Naturwissenschaften *12* (1924), 1137.

MILNER-BROWN, H.S., STEIN, R.B., YEMM, R. Changes in firing rate of human motor units during linearly changing voluntary contractions. J. Physiol. *230* (1973a), 371–390.

MILNER-BROWN, H.S., STEIN, R.B., YEMM, R. The orderly recruitment of human motor units during voluntary isometric contractions. J. Physiol. *230* (1973b), 359–370.

MILSUM, J.H. Control system aspects of muscular coordination. In: Medicine and Sport, vol. 6: Biomechanics II, pp. 62–71, Karger, Basel 1971.

MOMMAERTS, W.F.H.M. Stoichiometric and dynamic implications of the participation of actin and ATP in the contraction process. Biochem. biophys. Acta *7* (1951), 477.

MOMMAERTS, W.F.H.M. Muscular contraction. Physiol. Rev. *49* (1969), 427–508.

MORECKI, A., BUŚKO, Z.A., FIDELUS, K., KEDZIOR, K., OLSZEWSKI, J. Biomechanical modelling of dynamic properties of human motion. Proceed. IVth World Congr. Theory Machines and Mechanisms, Newcastle upon Tyne, Sept. 1975.

MORGAN, D.L. Separation of active and passive components of short-range stiffness of muscle. Am. J. Physiol. *232* (1977), C45–C49.

NATORI, R. Effects of Na and Ca ions on the excitability of isolated myofibrils. In: Molecular biology of muscular contraction (ed. Ebashi, S. et al.). Elsevier, Amsterdam, 1965.

NEEDHAM, D.M. Machina Carnis. Cambridge Univ. Press, 1971.

NORRIS, F.H. Active state plateau and latency in mammalian striated muscle. Am. J. Physiol. *200* (1961a), 667–671.

NORRIS, F.H. Isometric relaxation of striated muscle. Am. J. Physiol *201* (1961b), 403–407.

OTIS, A.B., FENN, W.O., RAHN, H. Mechanics of breathing in man. J. Appl. Physiol. *2* (1950), 592.

PARMLEY, W.W., SONNENBLICK, E.H. Series elasticity in heart muscle: Its relation to contractile element velocity and proposed muscle models. Circ. Res. *20* (1967), 112–123.

PAULI, W. Über den Zusammenhang von elektrischen, mechanischen und chemischen Vorgängen im Muskel. Kolloidchem. Beihefte *3* (1912), 361.

PEACHEY, L.D. Electrical events in the T-system of frog skeletal muscle. Cold Spring Harb. Symp. Quant, Biol. XXXVII (1973), 479–487.

PEACHY, L.D. The sarcoplasmic reticulum and transverse tubules of the frog's sartorius. J. Cell Biol. *25* (1965), 209–231.

PERRY, S.V., COTTERILL, J. Interaction of actin and myosin. Nature *206* (1965), 418.

PERRY, S.V., COLE, H.A., HEAD, J.F., WILSON, F.J. Localization and mode of action of the inhibitory protein component of the troponin complex. Cold Spring Harb. Symp. Quant. Biol. XXXVII (1973), 251–262.

PRYOR, M.G.M. Mechanical properties of fibres and muscles. Progr. Biophys. *1* (1950), 216.

RACK, P.M.H., WESTBURY, D.R. The effects of length and stimulus rate on tension in isometric cat soleus muscle. J. Physiol. *204* (1969), 443–460.

RAMSEY, R.W., STREET, S.F. The isometric length-tension diagram of isolated muscle fibres of the frog. J. cell. comp. Physiol. *15* (1940), 11–34.

RISEMAN, J., KIRKWOOD, J.G. Remarks on the physico-chemical mechanism of muscular contraction and relaxation. J. Am. Chem. Soc. *70* (1948), 2820.

RITCHIE, J.M. WILKIE, D.R. The dynamics of muscular contraction. J. Physiol. *143* (1958), 104–113.

SANDOW, A. Latency relaxation and a theory of muscular mechano-chemical coupling. Ann. N.Y. Acad. Sci. *47* (1946–47), 895.

SCHNEIDER, M.F. Linear electrical properties of the transverse tubules and surface membrane of skeletal muscle fibres. J. Gen. Physiol. *56* (1970), 640.

SIEGMAN, M.J., GORDON, A.R. Potentiation of contraction: effects of calcium and caffeine on active state. Am. J. Physiol. *222*( 1972), 1587–1593).

SNYDER, R.G. Link system of the human torso. Nation. Tech. Inform. Serv., US Dept. of Commerce, Springfield, 1972a.

SNYDER, R.G. Large deformation of isotropic biological tissue. J. Biomechanics 5 (1972b), 601–606.

SOONG, T.T., HUANG, W.N. A stochastic model for biological tissue elasticity in simple elongation. J. Biomechanics 6 (1973), 451–458.

STARK, L. Neurological control systems, p. 311. Plenum, New York, 1968.

STARLING, E.H., EVANS, C.L. Principles of human physiology. J. & A. Churchill Ltd., London, 1968.

STEVENS, J.C., DICKINSON, V., JONES, N.B. Mechanical properties of human skeletal muscle from *in vitro* studies of biopsies. Med. & Biol. Eng. & Comput. *18* (1980), 1–9.

SUGI, H. Tension changes during and after stretch in frog muscle fibres. J. Physiol. *225* (1972), 237–253.

SZENT-GYÖRGYI, A. Chemical Physiology of contraction in body and heart muscle. Acad. Press, New York, 1953.

THAMES, M.D., TEICHHOLZ, L.E., PODOLSKY, R.J. Ionic strength and the contraction kinetics of skinned muscle fibres. J. Gen. Physiol. *63* (1974), 509–530.

THOM, R. Structural stability and morphogenesis. Benjamin, Massachusetts, 1975.

TOMOVIĆ, R., BELLMAN, R. A systems approach to muscle control. Math. Biosci. *8* (1970), 265–277.

VARGA, L. The relation of temperature and muscular contraction. Hung. Acta Physiol. *1* (1946), 1.

VUKOBRATOVIČ, M., JURICIC, D. Contribution to the synthesis of biped gait. IEEE Trans. Biomed. Eng. *BME–16* (1969), January issue.

WALKER, S.M. Potentiation and hysteresis induced by stretch and subsequent release of papillary muscle of the dog. Am. J. Physiol. *198* (1960), 519–522.

WALSH, G.E. Physiology of the nervous system. Longmans, London, 1964.

WEBER, A., WEBER, H.H. Zur Thermodynamik der ATP-Kontraktion am Fasermodell. Z.f. Naturforsch. *5b* (1950), 124.

WEBER, A., WEBER, H.H. Zur Thermodynamik der Kontraktion des Fasermodells. Biochem. biophys. Acta. *7* (1951), 339.

WEBER, H.H. Das kolloidale Verhalten der Muskeleiweissköper. III. Physikochemische Konstanten des Myogens. Biochem. Z. *189* (1927), 407.

WEBER, H.H. Die Muskelkontraktion. Biochem. Z. *217* (1930), 430.

WENDT, I.R., GIBBS, C.L. Energy production of rat extensor digitorum longus muscle. Amer. J. Physiol. *224* (1973), 1081–1086.

WILKIE, D.R. Relation between force and velocity in human muscle. J. Physiol. *110* (1950), 249–280.

WILKIE, D.R. Measurement of the series elastic component at various times during a single muscle twitch. J. Physiol. *134* (1956a), 527–530.

WILKIE, D.R. The mechanical properties of muscle. Brit. med. Bull. *12* (1956b), 177–182.

WILLIAMS, W.J., EDWIN, A.I. An electronic muscle simulator for demonstration and neuromuscular systems modelling. Med. & biol. Eng. *8* (1970), 521–524.

WILSON, L.G. William Croone's theory of muscular contraction. Notes and records of the Royal Society of London *16* (1961), 158.

WÖHLISCH, E. Eine kolloidosmotische Hypothese der Muskelkontraktion. Verhandl. d. physikal.-med. Ges. Würzburg, Neue Folge *50* (1925), 163.

WOLEDGE, R.C. The thermoelastic effect of change of tension in active muscle. J. Physiol. *155* (1961), 187–208.

WOLEDGE, R.C. Heat production and chemical change in muscle. Progr. Biophys. Mol. Biol. *22* (1971), 39–74.

WOODS, H.J. The contribution of entropy to the elastic properties of keratin, myosin and some other high polymers. J. Coll. Sci. *1* (1946), 407.

YAMADA, H. Strength of biological materials. Williams & Wilkins, Baltimore, 1970.

# Index

A-band, 17, 18
Accelerometermyogram, 71
Acetylcholine transmitter substance, 39
Actin, 9, 11, 18, 100
Action potential, 32, 35, 37, 97
Activation heat rate, 136, 137
Active state, 13, 29, 32, 38, 40, 41, 97, 138
    definition of, 33
    of model muscle fibre, 104, 111
    of resting muscle (fibre), 34, 73
Actomyosin, 9, 137
ADP, 9
Analog computer simulation, 36, 37
ANDYMO (see Computer programs)
Anthropometric, 132
Anthropomorphic, 123
Applied generalized forces, 127
Asynchronous stimulation, 118, 119
ATPase, 8, 32, 57
Axon, 57

Bang-bang control, 14
Biceps muscle
    human, 145

Calcium ion, 9, 32
    concentration of, 33, 37, 38, 73, 97, 111, 151, 162
Collagen fibres, 23
Computer experiments, 96

Computer programs, 96
　ANDYMO, 131
　ELPEST, 83, 169, 171
　MYOSIM, 96, 179, 180
　SEMCI, 132
Computerized models
　of fibre, 98
　of muscle, 98
Configuration space, 125
Constant-volume relation, 10, 36, 112
Contractile
　element, 13, 19, 28
　force (externally observable), 31, 44
　proteins, 31
Contraction
　ballistic, 59, 73
　concentric, 97
　dynamics of muscle fibre, 42, 48, 98
　dynamics of whole muscle, 66, 150
　eccentric (see Eccentric contraction)
　isometric (see Isometric torque output)
　maximum effort, 79, 82, 91, 93, 95
　muscular, 1
　theories, 1, 5, 11
　times of motor units (see Motor unit)
Control
　model of muscle, 13, 64
　parametres, 14, 15, 39, 65, 89
Controlled-release method, 27
Controller subsystem, 123, 130
Cross-bridges, 9, 11, 28, 29, 100, 102
Cross-sectional area
　of fibre, 36, 58


Damping component, 12, 22
　of series elastic element, 27, 54
Dashpot, 15
Dehydration theory, 5
De-ionisation theory, 5
Dephosphorylation, 9
Depolarizing potential, 35
Distributed system, 51

Eccentric contraction, 89, 97
Elastic response, 25
Elastin, 24
Electromyogram, 71
Electromyographic recordings, 89, 90, 93, 94, 133
ELPEST (See Computer programs)
Endomysium, 16
Energy rate
    of muscular contraction, 136, 141
Entropy theories of muscular contraction, 6, 7
Epimysium, 16
Eulerian angles, 129
Excitation-contraction coupling, 33
Excitation dynamics
    of fibre, 31, 40, 42, 97, 118
    of whole muscle, 66
Excitation function, 140, 150, 163
    definition of, 64
Executor subsystem, 123, 125, 128, 130

Fascia, 18
Fasciculi, 16
Fibril, 17
Fibre (see Muscle fibre)
Filamentary-overlap function, 30, 72, 140
Filaments, 11
Force enhancement
    after stretch, 116
Force generator, 14, 15
Force-velocity relation, 43, 45
Fusiform muscle, 17, 56, 88, 129, 139

Globular heads, 9, 11, 29
Graphic display
    of model output, 96, 97

Half-relaxation time
    of fibre, 104, 107
H-band, 17, 18
Heavy-meromyosin, 9, 11, 29
Holonomic constraint, 127
Hominoid, 124, 125, 127, 131, 134
    definition of, 123

dynamics of, 122, 125
Humero-ulnar joint
human, 75
Hyperfast extensions, 43

I-band, 17, 18
Identification of model parameters (see Parameters)
Imbibition theory, 4
Inactive motor units, 58
Instantaneous centre of rotation, 75, 129
Intrafusal fibres, 12
Intrinsic strength
of fibre, 58
Inverse dynamic problem, 131
Isometric force, 45, 72
Isometric torque output, 75
linearly increasing, 91
quasi-stationary, 90
maximal dynamic, 79, 87
maximal steady-state, 79, 85, 144, 145
submaximal, 90

Lagrangian
of executor subsystem, 125
Latency, 105
Least-squares estimator, 77
of myodynamic parameters, 82, 89
Length-tension curve, 42
Length-tension relation, 32, 111
definition of, 42
for twitch contractions, 107
Lengthening (see Stretch)
Light-meromyosin, 9, 29
Long jump, 144
optimization problem of, 134, 135
take-off phase of, 130, 133, 134
Lumped system, 54

Maintenance heat rate, 136, 137
Maximum
isometric tension, 27, 58, 67, 72, 109, 118, 140, 146
performance, 144
shortening velocity, 44, 48

stimulation rate, 39
Maxwell model, 12
Metabolic energy
　of muscular contraction, 141, 142, 144, 145
Minimum-energy principle, 144
Model
　credibility of, 96
　definition of, 1
　validation of, 96
Models of muscle (see Muscle models)
Morphogenesis, 48
Morphology
　of muscles, 3
Morphometric parameters (see Parameters)
Motoneuron, 57, 123, 147
Motor unit, 57, 59, 97, 120, 147, 151
　contraction time, 61, 154
　maximum rate of derecruitment, 93
　maximum rate of recruitment, 65, 73, 93
　normalized speed of shortening, 61
　recruitment of, 15, 144, 146, 148, 162
　size law, 60, 148
　stimulation rate, 90
Muscle
　biomechanical models of, 12
　fibre, 16, 97
　mammalian, 1, 114, 118, 137
　models, 1, 13, 122
　moment arm function, 78, 81
　resting, 100
　skeletal, 1, 133
　spindle, 12
Myoactuators, 123, 127, 128, 129, 132, 141
Myoactuator subsystem, 129, 130
Myocybernetic
　control, 141, 142
　control model, 1, 56, 64, 96, 122, 134, 150, 179
　methods for parameter estimation (see Parameters)
　optimal control, 142, 143
　performance criterion, 141, 142, 143
Myocybernetics, 122, 136, 148
Myoenergetics, 136
Myodynamic methods (see Parameters)

Myodynamometer, 83, 91
MYOSIM (see Computer programs)
Myosin, 9, 18

Nerve, 1
  impulse, 38, 97
  potential, 35
Nervous system, 2, 141, 145
Neural control mode, 101, 109, 120
Neuro-musculoskeletal control system, 122, 130, 134, 142, 148
  definition of, 123
Normalized
  average stimulation rate, 65, 139
  calcium density, 36
  extension, 27, 72
  fibre force output, 97, 108, 110
  force-velocity curve, 46, 73, 74
  interfilamentary velocity, 44
  length of contractile element, 30, 66, 97, 150
  muscle force output, 97
  number of active motor units, 61, 97
  rate of motor unit recruitment, 65
  twitch force, 105, 106

Optimal control
  of long-jump take-off, 135
  of neuro-musculoskeletal system, 130, 142, 145
Optimality principles, 122, 136, 141, 142
Optimization of long jump (see Long jump)
Optimum length
  of contractile element, 44, 73, 165
  of muscle, 27, 72

Parallel elastic element, 12, 18, 21, 22
  of penniform muscle, 68
Parameters (methods of estimation)
  articular, 132
  morphometric, 71, 74, 132, 133
  myocybernetic, 72, 74, 93, 132, 133
  myodynamic, 71, 74, 79, 132, 133, 169
  of motor unit derecruitment, 95
  segmental, 132
  values of, 74, 86, 93

Penniform muscle, 17, 67, 72, 81, 87, 139, 165, 168
Performance criterion, 130, 136
　myocybernetic (see Myocybernetic)
Perimysium, 16
Polypeptide chains, 9
Posttetanic relaxation phase, 110
Principle of minimum transentropy, 147, 148
Principle of optimal grading sensitivity, 146, 148

Quadriceps femoris muscle group
　human, 92, 94, 137
Quasi-linear viscoelastic function, 25
Quick-release method, 27

Random sequence
　of torque measurements, 83
Recruitment dynamics, 66
Relative
　elongation, 22
　extension, 54
　facilitation, 113
　maximum stretching force, 45, 89
　stimulation rate, 39
　stretch potentiation, 113, 114
Relaxation phase, 111
Resting length
　of fibre, 102
Rheonomic constraints, 127

Sarcolemma, 16
Sarcomere, 18
Sarcoplasmic reticulum, 9, 32
Sarcoplasm, 29
Segmental parameters (see Parameters)
SEMCI (see Computer programs)
Semi-active motor units, 58, 159, 162
Sensory biosystem, 147
Series elastic element, 12, 18, 26
　methods for determining its constants, 27
　of cross bridge, 52
　of fibre model, 55
　of penniform muscle, 68

Shortening
    heat rate, 136, 139
    of fibres, 97, 117
    velocity, 139
Size principle, 58
Skeletal
    assemblage, 123, 125
    muscle (see Muscle)
Sliding-filament theory, 8
Spinal cord, 2, 147
State trajectory, 131, 134, 144
Stimulated motor units, 58, 97, 159
Stimulation rate, 15, 35, 97, 119, 137
    maximum of a motor unit, 62, 144
Stress-strain curve
    of fascia, 23
    of muscle, 23, 72
Stretch
    of fibre, 49, 97, 98, 111, 115, 117
    reflex, 89
Surface tension theory, 5
Switching function, 65, 161

Tendon, 1, 67, 74
    rest length of, 78
Tetanic tension (see Maximum isometric tension)
Tentanus, 111, 116
Terminal cisternae, 10, 29
Theories
    of muscular contraction, 2, 5, 8
Time-optimal control, 143
Transition process, 143, 144
Triceps muscle
    human, 75, 78, 81, 83, 85
Troponin, 33
Troponin-tropomyosin complex, 10, 11, 33
Transverse tubular system, 10, 32
Twitch contraction time
    of isolated fibre, 104
    of motor unit (see Motor unit)
Twitch response
    of fibre, 100, 104
Twitch-tetanus ratio, 104

Universal volume, 47

Velocity-dependence function, 31
  definition of, 44
Viscoelasticity, 24, 28, 127
Voigt model, 12

Work rate
  of muscular contraction, 136, 139

Z-disc, 21
Z-line, 17, 18